CROSSBILLS AND CONIFERS

CROSSBILLS AND CONIFERS

One Million Years of Adaptation and Coevolution

CRAIG W. BENKMAN

PELAGIC PUBLISHING

First published in 2025 by
Pelagic Publishing
20–22 Wenlock Road
London N1 7GU, UK

www.pelagicpublishing.com

Crossbills and Conifers: One Million Years of Adaptation and Coevolution

https://doi.org/10.53061/VSDN6841

A CIP record for this book is available from the British Library

ISBN 978-1-78427-552-5 Hbk
ISBN 978-1-78427-553-2 ePub
ISBN 978-1-78427-554-9 PDF

Typeset in ITC Stone Sans Std by S4Carlisle Publishing Services, Chennai, India
Printed and bound by CPI Group (UK) Ltd, Croydon CR0 4YY

EU Authorised Representative: Easy Access System Europe – Mustamäe
tee 50, 10621 Tallinn, Estonia, gpsr.requests@easproject.com

Cover photograph: Red Crossbill *Loxia curvirostra* foraging on
a Scots Pine *Pinus sylvestris* cone near Lake Ogallala, Nebraska
on 30 January 2024. © Steven G. Mlodinow

Frontispiece: Female Cassia Crossbill *Loxia sinesciuris* peering
perhaps at me, between episodes of probing for seeds between
the scales of a fallen Rocky Mountain Lodgepole Pine *Pinus contorta*
var. *latifolia* cone in the South Hills, Idaho on 4 June 2016.

5 4 3 2 1

Part of the art of becoming an ecologist ... is an ability to match the question one is asking with the most appropriate grouping of species—the set that will reveal valid patterns that act in nature.

(R.B. Root, quoted by A. Agrawal 2017: 151–2)

It affords me great pleasure to observe the rich and complex life of another species and to write its biography.

(G.B. Schaller 2012: 100)

Ponderosa Pine *Pinus ponderosa* cones, two of which were foraged on by Red Crossbills *Loxia curvirostra*. Rocky Mountain National Park, Colorado, 8 August 2019.

Contents

Preface

I was euphoric on my drive home. It was July 1997, and I had just spent the last week catching Red Crossbills *Loxia curvirostra* in the South Hills of southern Idaho. My last evening there was especially memorable, because I had seen the large numbers of crossbills I expected to encounter in an isolated mountain range where their main competitor for conifer seeds—the American Red Squirrel *Tamiasciurus hudsonicus*—was absent.

On my first visit to the South Hills the previous August, I gathered Rocky Mountain Lodgepole Pine *Pinus contorta* var. *latifolia* cones. I had flown to Salt Lake City, rented a car and was on my way to an ornithological conference in Boise. Based on the size and structure of the cones that I collected and my knowledge of crossbills and Lodgepole Pine to the east, on the other side of the Rocky Mountains, I thought it was likely that crossbills in the South Hills might be different. The crossbills I captured in the South Hills in 1997 had an average bill depth of 9.9 mm, which was near the 10 mm I expected. Although their average bill depth was only about 0.3 mm larger than the next-largest crossbill in the region, the difference was comparable to the differences between some of the other recognized forms of Red Crossbill in North America.

After I arrived home in Las Cruces, New Mexico, and viewed spectrograms of the contact calls I had recorded, I knew that the crossbills in the South Hills were distinct. However, a distinct call and a bill only 0.3 mm deeper than the next-largest crossbill in the region is meager evidence for proposing a new species of bird. Few bird species have been truly 'discovered' in the continental United States in the last century. Nearly all newly described species from this well-studied region and time period were the result of elevating to species previously described geographic races or subspecies that were already known to science as distinct forms.

No ornithologist had ever suggested anything special about the crossbills in the South Hills. Only the Gunnison Sage-Grouse *Centrocercus minimus*, which was recognized in 2000 (Young et al. 2000), represents a truly newly discovered species from the continental United States since 1919 (Peterson and Fisher 1955; Mayr 1957, 1971; Mayr and Vuilleumier 1983; Vuilleumier and Mayr 1987). Twenty years after capturing my first crossbill in the South Hills, this population of crossbills (which also occurs in the adjacent Albion Mountains) became the second newly discovered species in the last 100 years—the Cassia Crossbill *Loxia sinesciuris* (Chesser et al. 2017). But what really excited me on my initial drive home in 1997 was that I was convinced that a coevolutionary arms race between crossbills and Lodgepole Pine had led to the evolution of this distinct large-billed crossbill. In this book, I want to share what we have learned from Cassia Crossbills and crossbills more generally, including insights into coevolutionary arms races and diversification, cultural evolution and ecological speciation, and the form of adverse impacts of climate change. My goal is a book that will be fascinating to graduate students and academics interested in ornithology and evolutionary ecology, and to passionate naturalists and birders.

Crossbills are likely among the last choices for study. Or so it seemed for one of my graduate committee members—Ken Able—when we met to discuss my research plans. Ken was one of the most knowledgeable ornithologists in the region. He was incredulous and couldn't help but laugh when I first mentioned my idea of studying crossbills. I can still hear his chortle over 40 years later. Crossbills were uncommon in the northeast, where we were based at the State University of New York at Albany; even if seen, they were usually flying high overhead and were soon out of sight.

I had recently completed my MS thesis in northern Arizona, where I had studied Limber Pine *Pinus flexilis* with its large wingless seeds, and the foraging behavior of both its seed disperser, the Clark's Nutcracker *Nucifraga columbiana*, and its main seed predator, the American Red Squirrel. Like crossbills, both nutcrackers and Red Squirrels specialize on the seeds of conifers for their sustenance. The groundwork for the study of cone-bearing conifers and their cone-seed consumers was laid by Christopher Smith in his 1970 monograph on Red and Douglas Squirrels *T. douglasii* (together called pine squirrels) and Lodgepole Pine. Chris showed that it was straightforward to quantify the seed intake rates of these conifer-seed-eating squirrels simply by timing how long they take to remove seeds from the cones, then husk and swallow the seed kernels. By measuring the masses and calories of the seed kernels, he was able to estimate their energy intake while foraging.

I already knew that crossbill feeding behavior atop trees could be similarly quantified using a telescope and stopwatch. Because crossbills are so specialized for foraging on seeds in conifer cones, it was also easy to figure out their diet. Data on consumer–resource interactions are fundamental for much of ecology, but quantifying diet accurately for the vast majority of organisms is usually not a straightforward task (Pringle and Hutchinson 2020). Crossbills are also tame and cooperative in captivity and so by gathering cones for them to feed on, I knew I could ask complementary questions to those I could study in the wild. Given that measuring seed intake rates (i.e., resource profitability) was instrumental to exploring many of the topical questions in behavioral ecology and ecology in the early 1980s when I was a graduate student, crossbills and conifers seemed superior to most study systems for which resource availability was a black box.

The preeminent ecologist G. Evelyn Hutchinson wrote thus in one of his classic papers:

> In any study of evolutionary ecology, food relations appear as one of the most important aspects of the system of animate nature. There is quite obviously much more to living communities than the raw dictum 'eat or be eaten,' but in order to understand the higher intricacies of any ecological system, it is most easy to start from this crudely simple point of view.
>
> (Hutchinson 1959: 147)

Following Hutchinson's lead, I focus the first three chapters of this book on the consumer–resource interactions or 'food relations' between crossbills and conifers. Extracting seeds securely wedged between closed and woody cone scales is literally a tough nut to crack! Crossbills have solved it with the evolution of a novel adaptation from which they get their name. Crossed mandibles have opened up many opportunities but also come at a cost to them. The evolution of the mandible crossing and its consequences are the focus of Chapter 1. One of the challenges of specializing on seeds in conifer cones is that the annual seed crops that most crossbills rely upon vary tremendously in size over space and time. Such spatio-temporal variation is an effective deterrent for most seed predators. I focus on some of the behavioral responses of crossbills to this variation in Chapter 2, and explore how subtle differences in both cone and bill structure influence crossbill behavior and ecology in Chapter 3.

Still following Hutchinson, I build upon the foundation in the first three chapters of 'food relations ... to understand the higher intricacies'. This includes some of the foremost questions in evolutionary ecology. In Chapter 4, I ask which resource characteristics allow, if not favor, specialization, and how the diversity of conifers with such traits has influenced the diversity of crossbills. In Chapter 5, I address how the interactions between crossbills and conifers drive reciprocal selection that has resulted in coevolutionary arms races and diversification. Like many evolutionary biologists before me, I have gained much insight by studying birds on both islands and 'sky islands' such as those occupied by Cassia Crossbills in southern Idaho, and by comparing them to populations elsewhere. Chapters 6, 7 and 8 show how reproductive isolation arises between the diverging lineages. I focus on crossbill flocking behavior in Chapter 6. I became interested in flocking behavior because my field observations of foraging crossbills suggested that they used the feeding performance and contact calls of flockmates to help assess tree quality (i.e., seed intake rates). If this were true, which seems to be the case, then crossbills should flock assortatively by feeding ability or bill size. This in turn could lead to assortative mating, because crossbills flock year-round and likely choose mates from within their habitual flocks. I discuss our experiments to test these ideas in Chapter 6 and I examine more broadly the ecology of crossbill speciation in Chapters 7 and 8. Threats to crossbills are the subject of Chapter 9, focusing mostly on climate change. Many species suffer from climate change via consumer–resource (trophic) interactions, and knowledge of such interactions for crossbills has been critical for elucidating and anticipating impacts on them. In the final chapter, Chapter 10, I discuss some general implications of our research, as well as some outstanding questions that still puzzle me.

Many people have influenced my research career along the way, several of whom were instrumental in my decision to study crossbills. My master's thesis advisor, Russell Balda, suggested the project on Limber Pine, nutcrackers and Red Squirrels. I was primed, having recently read Chris Smith's classic monograph on coevolution between conifers and pine squirrels (Smith 1970)—a friend from undergraduate school told me that his mammalogy instructor recommended it highly. It took little to persuade me to work on the project that Russ had devised with Chris the previous winter. Within minutes after discussing the project during my first day on campus in August 1979, Russ drove me to my future study site on the west slope of the San Francisco Peaks just north of Flagstaff, Arizona.

After finishing my master's degree, I moved to Albany, New York where I did my PhD, working with H. Ronald Pulliam. Ron was the first to reveal to me the utility of studying seed-eating birds when I attended his lecture at my first ornithological conference in the summer of 1977. In his lecture, Ron demonstrated that he could predict rather well the relative abundances of the different species of sparrows in the grasslands of southern Arizona by combining measurements of the intake rates of captive sparrows on different species of seeds with the abundances of the different seeds (Pulliam 1975). I was excited by the ideas permeating the field in both foraging theory, of which Ron was one of the architects, and how competition influenced community structure. Few systems appeared to allow such straightforward integration, but seed-eating birds seemed to be one. Ron further emphasized the value of studying a system where you could combine fieldwork with aviary experiments.

When I started my PhD, I initially planned to study sparrows and the seeds they eat, which is what Ron knew so well. But I was discouraged by botanists at the state museum, who were dubious about the idea that some seed traits might have evolved as defenses against seed predators. I just wasn't familiar enough with sparrows and grasses, whereas after hiking hundreds of miles in western conifer forests and familiarizing myself with conifers and the literature, I was more comfortable with conducting research on crossbills. Ron was supportive. So, crossbills it would be. Jeff Brawn, whom I met while registering for classes during my first semester of graduate school, likes to remind me that it was at his field site in the Ponderosa Pine *Pinus ponderosa* forests south of Flagstaff where I first talked about studying crossbills. It was about a half a year later that I convinced myself that I should try. I never anticipated I would spend the next 40 years studying them.

Craig W. Benkman, April 2025

Acknowledgments

I have tried to mention in the text those who have been instrumental in the research I discuss in this book. Nevertheless, I still might not have recognized adequately their contributions. For example, I mentioned my PhD advisor, Ron Pulliam, in the Preface but I do not mention him in Chapter 1. Yet I am pretty certain that, while driving with Ron to his field site at the E.N. Huyck Preserve south of Albany, New York, he suggested removing the mandible crossing as an experiment to determine its adaptive significance, an experiment I describe in Chapter 1. When I mentioned this to Ron several years ago, he emphasized that we should simply appreciate that ideas often arise from interactions. I hope I haven't underestimated such interactions in this book.

Cody Porter read and provided many helpful suggestions on two drafts of the book as I completed them, and helped convince me that this book was worth sharing with others. John N. Thompson read a subsequent draft of the book and provided numerous thoughtful suggestions and encouragement, as he has done for over 30 years. Eliot Miller read most of the book and noted especially those sections in need of revision. Pim Edelaar and Ethan Linck read and commented on the whole book and made many useful suggestions to improve its presentation. Ethan in particular made countless stylistic suggestions that made the manuscript more readable. Irby Lovette read the first five chapters, and most paragraphs had helpful suggestions from him for improving the presentation. I owe Irby a special thanks for encouraging me to write this book. Trevor Price also made the same suggestion and has been quite supportive over the years, including even mailing my graduate student Julie Smith a computer to help with certain analyses when I had limited research funds. Bob Montgomerie read the Preface and the first part of Chapter 1, providing recommendations on writing style. Ted Floyd made clear that work was needed for the book to appeal more broadly.

Multiple curators and collections managers have either provided access to crossbill specimens or loaned them to the University of Wyoming's Museum of Vertebrates where I could measure them. The curators and collections managers directly responsible for aiding in the measurements presented in this book include Paul R. Sweet and Peter Capainolo, American Museum of Natural History; Maureen Flannery and Laura Wilkinson, California Academy of Sciences; Stephen P. Rogers, Carnegie Museum of Natural History; Ben Marks, Field Museum of Natural History;

John McCormack, Moore Laboratory of Zoology; J.V. Remsen and Steve Cardiff, Museum of Natural Science; Carla Cicero, Museum of Vertebrate Zoology; Robert Prys-Jones, Natural History Museum, Tring; Kristof Zyskowski, Peabody Museum of Natural History; Philip Unitt, San Diego Natural History Museum; James Dean, Jacob R. Saucier and Brian K. Schmidt, Smithsonian Institution; Nate Rice, The Academy of Natural Sciences of Drexel University; Kathy C. Molina, UCLA-Dickey Bird and Mammal Collections; Mark B. Robbins, University of Kansas Biodiversity Institute; René Corado, Western Foundation of Vertebrate Zoology.

I only mention Alan Knox briefly in the text, but he graciously hosted me in Aberdeen and over the years repeatedly responded to my queries on crossbills and locations for finding them.

I am especially thankful to the many photographers who kindly let me use their photographs, including David Whitaker who rescanned old slides, and to Vanessa Powell at The Cornell Lab of Ornithology, Macaulay Library who made the initial contact with most of the photographers. David Allen Sibley kindly allowed me to use his illustrations in Figures 1.2 and 1.5.

I thank Oxford University Press, John Wiley and Sons, *The Wilson Journal of Ornithology*, Royal Society of London and *The Forestry Chronicle* for allowing me to reproduce figures, Bloomsbury Publishing for allowing me to use text and The University of Chicago Press and Elsevier for allowing to me to reproduce figures and use text in Chapters 2, 4 and 9 that first appeared in Benkman (2016) and in a chapter in a book (Benkman 2024), respectively. I thank Nigel Massen, the Publisher at Pelagic Publishing, for encouraging my submission, and David Hawkins, the Production Editor, and Sara Magness, the Copy Editor, for guiding and further editing my manuscript.

Financial support for the research discussed in this book has come from the Frank Chapman Fund, the National Science Foundation, NSERC of Canada, the National Geographic Society, New Mexico State University, University of Wyoming, and Robert B. Berry who provided generous funds to the University of Wyoming Foundation, which both endowed my faculty position and provided research funds that were especially helpful when I was unsuccessful acquiring funds elsewhere.

Finally, I thank my wife Jennifer who for 35 years has shared my love of nature and has been supportive of my obsession with crossbills.

Male Cassia Crossbill *Loxia sinesciuris* in the South Hills, Idaho, 10 June 2017.

Chapter 1

Why Crossbills? Crossbills, Conifers and the Origins of an Interaction

> Were I an ornithologist, I think that the finches of the genus *Loxia* would take up most of my research time. No group of birds seems to offer more tantalizing problems in that area of biology where systematics, ecology, zoogeography, population dynamics, and ethology overlap.
>
> (W.L. Brown Jr. 1957: 267)

Although Bill Brown's sentiments help explain why I continue to study crossbills *Loxia* spp., I was initially drawn to them because I knew it would be straightforward to measure rates at which they ate seeds. Crossbills feed nearly exclusively on seeds in conifer cones, mostly located near the outer tips of branches, readily visible to someone with binoculars or a telescope. As the crossbills use the same general behaviors or stereotypic movements to extract and husk seeds (Fig. 1.1),[1] I was able to quantify how many seeds a foraging crossbill eats per unit time.

I first watched foraging crossbills on a three-day 'Birds of Yosemite' field course in June 1975. A family friend who was a casual birder recommended the course, knowing my love for the Sierra Nevada and my budding interest in its natural history. She also knew of the inspiring instructor, Dave DeSante, whose passion and knowledge got me hooked on birds. On the final day, we watched a lone Red Crossbill *Loxia curvirostra* foraging on a fallen Lodgepole Pine *Pinus contorta* cone as our group encircled the bird. I could see right away both the ease at which foraging crossbills could be observed (Fig. 1.1) and that they seemed to be undisturbed by our close presence.

A few years later I began to appreciate the utility of being able to observe crossbills so intimately. Here, Chris Smith's pioneering work on pine squirrels

1 For a video, see  or https://www.youtube.com/watch?v=1NvU8WG9bg0

FIGURE 1.1 Female Cassia Crossbill *Loxia sinesciuris* lifting a seed from between the scales of a Rocky Mountain Lodgepole Pine *Pinus contorta* var. *latifolia* cone with her tongue (A) and raising her head (B) while she removes the seed coat (husks the seed) and swallows the kernel in the South Hills, Idaho on 4 June 2016. These motions are distinctive, including the rapid lateral motions of the lower mandible while husking seeds (these lateral motions are not used for empty seeds), enabling researchers to record how many seeds are successfully extracted and eaten.

Tamiasciurus spp. and conifers (Smith 1968, 1970; see epigraph of Chapter 5) was instrumental. Chris showed that the mass of the individual seed kernels within a conifer species is among the least variable of conifer reproductive traits. A modest sample of seeds from a relatively small number of trees suffices to characterize average kernel mass. Thus, as Chris had done, I could convert the number of seeds consumed into kernel mass consumed by weighing the seed kernels. Conifer kernels also tend to vary little in food composition (i.e., protein, fat, specific caloric value), allowing me to use measures from the literature to convert mass of kernel to calories. Not often is the energetic content of food consumed so easily measured, and rarely is it a sufficient measure of resource value because quality generally varies among foods. Quality varies because prey might differ in their size, their nutrient composition and in their chemical defenses. Plants are known for their diversity of chemical defenses that make them difficult to digest and even harmful if not fatal for consumers.

Crossbills do not face this kind of chemical challenge, as the seed defenses of most conifers they feed upon are primarily structural, in the form of the woody cone. Thus, I realized that I could simply use the time it takes a crossbill to extract and husk a seed as the 'cost', and kernel mass as a measure of the 'benefit'. Such a simple means of measuring feeding (energy) intake rates was appealing to me because it allowed me to focus on the behavior and ecology of crossbills, which were my primary interest. Feeding intake rates were particularly useful because they are a direct measure of food profitability, which was foundational to the theories of foraging behavior and species coexistence that were of intense interest when I was a graduate student. The model study systems for testing such theories in animals were those where feeding intake rates could be measured either in nature (e.g., Gill and Wolf 1977; Grant 1986) or in the laboratory in a manner meaningful to nature (e.g., Pulliam 1975; Werner 1977).

Conveniently, crossbills are a delight in captivity. They rapidly adjust, and within minutes of capture forage for seeds in cones even when I am but a few feet away. Moreover, conifer cones of many different types could easily be gathered and presented to the captive birds, where I could observe foraging behavior just as in the field. I could bring in not just the foods but also the foraging substrate (i.e., seeds in cones on branches). This gave me more options and control than if I could only follow the birds in the wild. To ensure that feeding performance in aviaries was similar to that in the wild, we gathered vast numbers of cones to give to the birds in the aviaries. We could use cones from different species or alter their ripeness, change the number of seeds in them, and even modify the presumed defenses such as spines on the scales by trimming them. These experiments allowed us to understand how both cone features and the phenology (seasonality and timing) of cone ripening affect crossbill intake rates. We could also capture crossbills of different sizes to determine how bill size affects performance on different types of cones. In short, the ability to measure costs and benefits, and how they varied in relation to variation in both crossbills and conifers, made crossbills highly tractable for understanding many of the aspects of their behavior, ecology and evolution that Bill Brown alluded to in the epigraph of this chapter. In this chapter, I focus on how crossbills access seeds in cones, on their adaptations for extracting seeds from cones and on some of the consequences of such specialization.

Why crossed bills?

To understand how crossbills use their uniquely shaped bills to extract seeds from tough cones, we start with an unusual feature of their foraging anatomy: crossbills (like many other finches in the subfamily Carduelinae) have the ability to move their jaw (lower mandible) laterally while husking seeds. In contrast, most seed-eating birds primarily use their mandibles in the vertical dimension, like simple pliers, as they crush seeds through their biting forces. Cardueline finches have various muscular and skeletal features plus an absence of the postorbital ligament, which collectively enables them to move their lower mandible to the side (mediolateral movement or abduction) in addition to the vertical movements used by other seed-eating birds (Nuijens and Zweers 1997; Nuijens and Bout 1998). This greater range of motions increases the angles and potential forces exerted on seeds, which likely enables crossbills and their relatives to husk a diversity of seed types efficiently (van der Meij and Bout 2006; Mielke and Wassenbergh 2022).

This foraging flexibility may have been instrumental in the cardueline adaptive radiation into about 190 species of mostly seed-eating birds with a wide range of bill sizes and shapes (Fig. 1.2; Nuijens and Zweers 1997). In addition to crossbills, carduelines include the familiar Evening Grosbeak *Hesperiphona vespertina* and Pine

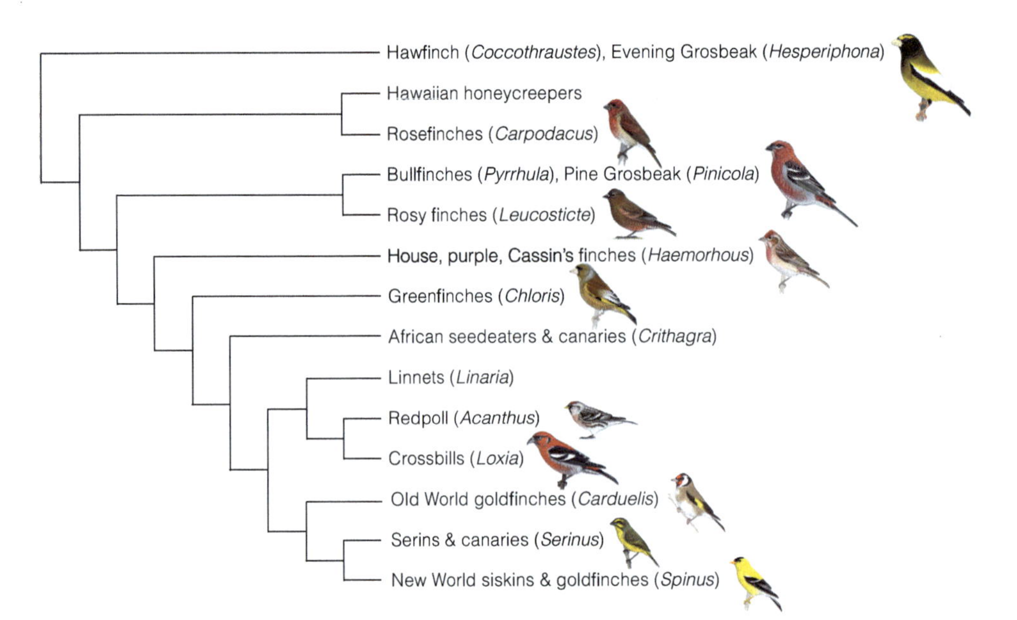

FIGURE 1.2 A phylogeny or diagram of evolutionary relationships among many of the well-known lineages within the subfamily Carduelinae, labeled by the common and generic names of representatives (modified from Zuccon et al. 2012). Crossbills are most closely related to Redpolls *Acanthus flammea*; their most recent common ancestor is estimated to have occurred between 750,000 and 1.4 million years ago (C. Brock, pers. comm. 2023), which is substantially less than a recent estimate of over 4 million years ago (Päckert et al. 2020). The birds illustrated are males, scaled in size, of the following species from top to bottom: Evening Grosbeak *Hesperiphona vespertina*, Common Rosefinch *Carpodacus erythrinus*, Pine Grosbeak *Pinicola enucleator*, Brown-capped Rosy Finch *Leucosticte australis*, Cassin's Finch *Haemorhous cassinii*, Oriental Greenfinch *Chloris sinica*, Redpoll, White-winged Crossbill *Loxia l. leucoptera*, European Goldfinch *Carduelis carduelis*, Yellow-fronted Canary *Serinus mozambicus*, American Goldfinch *Spinus tristis*. Bird illustrations from Sibley (2014) with permission from David Allen Sibley.

Grosbeak *Pinicola enucleator*, bullfinches, House Finches *Haemorhous mexicanus*, greenfinches, canaries, linnets, Redpoll *Acanthus flammea*, goldfinches and siskins. The exceptionally diverse Hawaiian honeycreepers are also cardueline (Fig. 1.2).

The ability to laterally abduct their lower mandibles allows some cardueline finches to access seeds in conifer cones. I've watched Pine Siskins *Spinus pinus* and Redpolls abduct their lower mandibles to spread apart the small, thin cone scales of White Spruce *Picea glauca* and Tamarack *Larix laricina* (see Fig. 1.9B) to reach the underlying seeds. However, siskins and Redpolls only spread apart thin scales that have previously opened. Apparently, they cannot create a gap between cone scales that haven't dried and separated at least slightly. Similarly, I've watched a Hawaiian honeycreeper ('Akeke'e *Loxops caeruleirostris*) laterally abduct its lower mandible to access insects between the overlapping leaves in leaf buds, a structure similar to, albeit less woody than, the scales of conifer cones (Benkman 1989b).

Redpolls are the sister taxa to crossbills (Fig. 1.2; i.e., they are more closely related to each other than to any other taxa), and it is easy to envision a redpoll-like ancestor exploiting seeds from between the thin, papery scales of White Spruce or Tamarack in the northern boreal forests in winter, much like Redpolls do now. It is also easy to envision that in years when the readily accessible seeds have been eaten or have fallen from the cones, any individual that could access seeds between closed cone scales would have an advantage. Crossed mandibles (Fig. 1.3) provide such an advantage by enabling the tips of the bill to spread apart closed scales. The slender, curved mandible tips are especially useful, because they allow crossbills to both aim their biting forces precisely and reduce the forces needed to slide between the woody scales. More power can be provided at the tip of the mandibles because with increasing curvature, the forces a mandible experiences switch from shearing to compression, reducing the risk of mandible fracture (Bowman 1961). Think of your teeth. If your teeth stuck straight out of your mouth and you attempted to bite hard on a solid object, your teeth would either shift in their sockets or fracture. Neither is desirable. Thankfully, and for good reason, our teeth are vertical and opposing so that when we bite, our teeth experience compression rather than shearing, allowing us to exert and withstand greater forces. The same should hold true for crossbills.

FIGURE 1.3 The lower mandible twists to the side to cross the upper mandible, while the hook of the upper mandible, like the hook or overhang found at the tip of the upper mandible in many bird species (Box 1.1), does not twist to the side. When I refer to mandible crossing direction, the direction is the side to which the lower mandible crosses; to its left in this male Cassia Crossbill *Loxia sinesciuris* photographed in the South Hills, Idaho on 9 June 2017.

Box 1.1: Bill hooks

Many readers might wonder about the hook that occurs at the tip of the upper mandible of numerous bird species. Hooks found in raptorial birds like falcons, hawks and owls are pronounced and have an obvious function. They tear flesh. For the majority of bird species, bill hooks are less obvious, if present, and so are the benefits. The songbirds with among the most pronounced hooks are neotropical flowerpiercers *Diglossa* spp. As their name implies, flowerpiercers use their bills to pierce tubular flowers at the base to access nectar that is otherwise inaccessible except to birds with long, slender bills—like hummingbirds—that can reach the nectar through the corolla opening. By clipping the hooks off the upper mandibles of Cinnamon-bellied Flowerpiercers *D. baritula*, Jorge Schondube and Carlos Martinez del Rio (2003) found that the hook aids the birds by securing the corolla while the straight lower mandible pierces it from the side. And as I discuss at the end of this chapter, there is a cost to specialization: the hook impedes their ability to handle fruit, the flowerpiercers' main alternative food.

A more widespread benefit of a hook or overhang at the tip of the upper mandible appears to lie in aiding ectoparasite removal (Clayton and Walther 2001; Freed et al. 2008; Clayton et al. 2010). Dale Clayton and colleagues used various ingenious techniques to reveal that the overhang in Rock Pigeons *Columba livia* provides a surface against which the lower mandible creates destructive sheering forces, killing feather lice (Clayton et al. 2005). Without the overhang, feather lice increase in abundance to the pigeon's detriment. The importance of well-occluded mandibles for preening is why Dale suspects crossbills, whose mandible tips are anything but well-occluded (Fig. 1.3), appear to have elevated levels of ectoparasites (Pomeroy 1962). Unfortunately, Dale and I have yet to coordinate a field trip to test for such a pattern.

The naturalist and ornithologist Desmond Nethersole-Thompson envisioned a scenario for the evolution of crossbills in his book, *The Pine Crossbills*:

> Their common ancestor was presumably a small cone-seed eating finch with a straight beak. Selection later perpetuated a new stock bred from a few individuals whose beaks tended to cross. With its crossed mandibles the emergent crossbill was then able to extract seeds from closed cones and thus eventually to breed earlier than the parent stock which could only exploit opening or discharging cones.
>
> (1975: 21–2 © D. Nethersole-Thompson, 1975, *The Pine Crossbills*, T & A D Poyser, an imprint of Bloomsbury Publishing Plc)

A genetic basis for crossed mandibles, necessary for this scenario, has been found in various breeds of chicken (Landauer 1938; Bai et al. 2018; Joller et al. 2018); although bill 'deformities' need not always have a genetic basis (Zylberberg et al. 2018). Moreover, it is not a stretch to envision a slight mandible crossing arising in an ancestor of crossbills, given that mandibles that cross at the tip are an uncommon variant among numerous bird species (Pomeroy 1962). My own first sighting of such a bird was a gull along the coast of California soon after I started birding in 1975. As with this gull, the mandibles of an ancestral crossbill likely crossed only slightly.

But would a slight crossing provide access to seeds that were not readily available to those with straight mandibles? And would further increases in the crossing be even more beneficial? If so, then it would be easy to envision how mutations that led to a slight crossing would have been favored. And mutations that led to further crossing would have also been favored. While the natural world abounds with situations like this, where it is easy to posit how evolution could have favored an unusual trait, the beauty with crossbills and conifers is that we can use them to experimentally test Nethersole-Thompson's scenario about the adaptive benefit of having crossed mandibles. The crossing is all rhamphotheca (Newton 1972)—the outer horny covering of the bill that grows and wears much like our fingernails. And variation in wear depending on the accessibility of seeds creates birds that are at these different stages, from nearly uncrossed to highly crossed.

Anna Lindholm helped conduct an experiment to test this scenario for the evolution of the mandible crossing (Benkman and Lindholm 1991). Anna was then an undergraduate at the University of British Columbia, where I was a postdoctoral fellow. She would go on to become an accomplished scientist, now on the faculty at the University of Zurich. Inspired by her vertebrate biology course, Anna stopped by my office one day, keen on getting involved in a research project. At that time, I had been conducting aviary experiments using four forms of Red Crossbill common in the Pacific Northwest; these experiments will be discussed in Chapter 4. These four taxa differ in average bill size and are categorized by their contact call as 'call types'. This categorization was developed by Jeff Groth who at the time was a graduate student at the University of California at Berkeley (Groth 1993a).

Anna and I chose to use the smallest of these call types (Type 3—Jeff numbered them in the order he discovered them; this call type corresponds to the previously recognized subspecies *L. c. minor*) for our experiment, for several reasons. First, we had a large number of the green, closed Western Hemlock *Tsuga heterophylla* cones on which these small crossbills are specialized (Fig. 1.4; Benkman 1993a). In addition, we could produce open cones by placing green cones on the floor to dry, and could easily reclose them by misting them with water. Cone reclosing occurs commonly in humid conditions and is important to crossbills because the reclosing helps retain the seeds in the cones. Finally, the thin scales of Western

FIGURE 1.4 A male Red Crossbill *Loxia curvirostra* (most likely Type 3) biting to separate the closed scales of a thin-scaled Western Hemlock *Tsuga heterophylla* cone. This crossbill's lower mandible crosses to its left and it is holding the cone with its right foot. Type 3 crossbills are the smallest of the New World Red Crossbills and are adapted for foraging for seeds in Western Hemlock cones. Photograph taken by John Riegsecker at the Theler Wetlands in Washington State on 2 September 2020.

Hemlock cones (Fig. 1.4) are similar to the cones that Redpoll commonly forage upon, and the Type 3 crossbills are among the smallest of all crossbills (although still approximately double the mass of a Redpoll).

Anna first measured the amount of time that each of seven hungry crossbills (they had all fasted overnight) required to remove and husk 10 seeds. She timed each crossbill foraging on 10 closed hemlock cones, 10 reclosed cones and 10 open cones. Then I used nail clippers to remove the mandible crossing from four of the crossbills under the watchful and curious eyes of the university veterinarian. I tried to trim the mandibles so that they were like those of an average non-crossbill finch, although the tip of the mandibles still crossed slightly when I was done. It was hard to remove all the rough edges, so I provided grit-covered perches and smeared pitch on the bills to encourage the crossbills to wipe their bills to further smooth the cut edges. Then as before, Anna measured the intake rates of all seven crossbills including the three unaltered crossbills that served as controls. Anna repeated these measurements every few days for five weeks as the mandibles of the modified 'uncrossed' crossbills grew back to their original lengths.

The results were striking. After removing nearly all the crossing, none of the four birds with trimmed mandibles could extract seeds from closed cones. Only after the crossed part of the mandible reached about half its original length was a crossbill again able to take seeds from closed cones. As the mandibles continued to grow and the crossing increased, the time to extract seeds from closed cones decreased, eventually converging on the feeding times of the unmodified birds. This gradual increase in feeding performance is unlikely to be attributed simply to recovery from an injury, because individuals with trimmed mandibles had intake rates on open cones that were similar to both those before trimming and those of the controls. In the case of reclosed cones, the modified crossbills were unable to access seeds right after their crossings were trimmed. But after a few days of regrowth and a slight increase in mandible crossing, the crossbills could again extract seeds. And as with the closed cones, the times to extract seeds from reclosed cones converged on their pre-trimming times as their mandibles approached their original lengths. This suggests that during the incipient evolution of the crossed bill, a slight crossing would indeed have provided access to seeds that were unavailable to birds without a crossed bill, and that further increases in mandible crossing could have been favored because this would have provided ever-greater access to seeds, including those in closed cones.

Genomic data (Chad Brock, pers. comm. 2023) suggest that the earliest split among extant crossbills divided the White-winged/Hispaniolan Crossbills from the Two-barred and Red Crossbills, and that this earliest division occurred perhaps 500,000 to 2.5 million years ago (Fig. 1.5). As the Two-barred, White-winged and Hispaniolan Crossbills collectively share a set of 'wing-barred' traits, it is likely that these same traits were present in the shared crossbill ancestor. It seems likely that the ancestor of the Red Crossbills split from their closest relative,

FIGURE 1.5 A phylogeny of crossbills showing the main and most strongly supported splits. A male Red Crossbill *Loxia curvirostra* from North America and a male White-winged Crossbill *Loxia l. leucoptera* are shown to scale. All 'Red Crossbill' taxa are fairly similar in plumage whereas they differ substantially in bill and body size. The three lineages of wing-barred crossbills (Two-barred, White-winged and Hispaniolan) are similar in plumage, and each is represented by a single species or subspecies. The difference between the average bill depths of the smallest and largest taxa is substantially smaller in the wing-barred crossbills (2.30 mm) and New World 'Red Crossbills' (2.80 mm) than in the Old World 'Red Crossbills' (5.92 mm). Bird illustrations from Sibley (2014) with permission from David Allen Sibley.

the Two-barred Crossbill, probably in Eurasia since that is where the Two-barred Crossbill is found. Red Crossbills may have then colonized North America about 160,000 to 550,000 years ago. Alternatively, the split between Two-barred and Reds could have occurred by an initial colonization of North America and then a subsequent colonization of Eurasia by Reds. Regardless of which scenario is correct, independent radiations in each hemisphere resulted in a current total of 26 'Red Crossbill' taxa (Fig. 1.5).

Getting seeds from cones

The mechanics of how crossbills forage on closed and reclosed cones have long fascinated ornithologists, and have been well described by Tordoff (1954) and Newton (1972). But in addition to the novelty of how crossbills extract seeds from cones, knowing how crossbills forage is key for understanding the evolutionary interactions between crossbills and conifers. Their foraging behavior determines which cone traits act to deter crossbill feeding, as well as which bill traits act to counter seed defenses.

When I first started studying captive crossbills, I watched them through a one-way window from only about a foot away (Fig. 1.6). This enabled me to observe crossbills foraging in great detail. I never tired of watching crossbills, even after seeing them eat many thousands of seeds. It is a marvel to watch crossbills maneuver, including hanging upside down at the end of branches (see cover photograph) to orient and then adeptly using their bills to spread apart closed

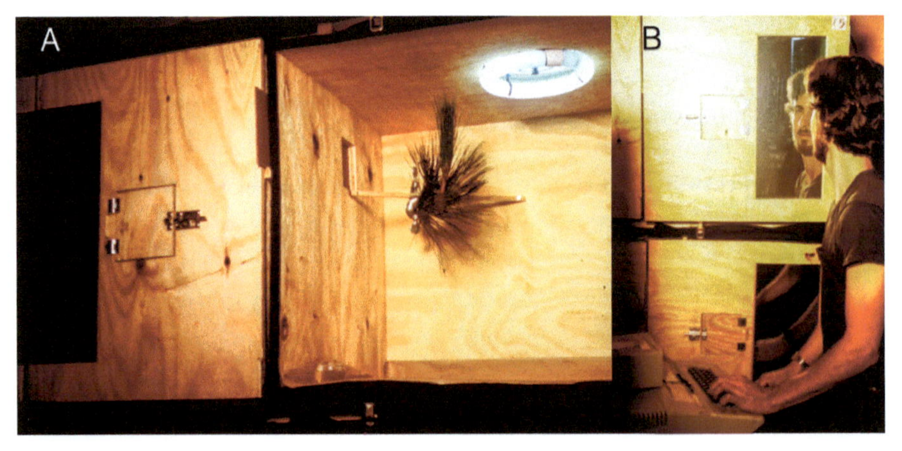

FIGURE 1.6 The foraging box setup used during my dissertation. (A) A crossbill was placed in the left box in the evening with the lights set to a timer. The following morning, I secured a branch with a cone attached in the right box. Then I turned off the lights, opened the small sliding door separating the two boxes and then turned on the light in the right box so that the crossbill would enter and forage on the cone, while (B) I watched through a one-way window and recorded foraging behavior with an Apple computer programmed as an event recorder.

cone scales to reach the underlying seeds. More than once while watching through the one-way window, I had to catch myself from clapping in admiration at what these feeding birds could accomplish.

Foraging crossbills perch on or near the cone, and, especially when the cone is loose, secure it with the claws of one or both feet, depending on their perch and size of cone. Crossbills have sharp claws, strong feet and sturdy legs that enable them to hold cones securely and climb acrobatically, a bit like parrots. Seeds—usually two per scale—are located at the base of overlapping scales that resemble shingles on a roof and spiral around the cone. To extract seeds, crossbills open their bill, then bite between the overlapping scales with the tips of their mandibles to form a gap (see Figs 1.4 and 1.7A). Crossbills orient themselves so that the side to which the lower mandible crosses is on the side toward the center of the cone (on the birds' right in Figs 1.7B,C, and on the bird's left in Fig. 1.4). Once the mandible tips reach between the scales to provide sufficient leverage, crossbills spread apart the scales by moving (abducting) the lower mandible to the side to which it is crossed, while the upper mandible pushes the scale on the other side of the gap in the opposite direction (Fig. 1.7B,C). If this does not expose a seed, the mandibles probe deeper into the gap and then the lower mandible is abducted until a seed can be lifted free by the bird's protrusible tongue (Fig. 1.7C). Seeds seem to stick to the moist surface of the tongue by capillary adhesion, just as small seeds adhere to my wetted finger. Much like in hummingbirds and woodpeckers, the bony support of the tongue—the hyoid apparatus—extends to the back of the skull, although not as far as in woodpeckers. This enables crossbills to extend their tongues into the gap between scales to lift the seeds free. Sometimes crossbills also use their upper mandibles to hook seeds free from the base of the scales.

An extracted seed is moved with the aid of the tongue toward the base of the bill. Here the tongue secures the seed in a lateral groove in the horny palate of the upper mandible (see Fig. 4.6) while the sharp edge of the lower mandible (tomium) cracks the seed coat (see Fig. 4.3A), which is removed and discarded by movements of the tongue and lower mandible. Then the kernel is swallowed. Because of the asymmetry of the mandible crossing, seeds are secured in the lateral groove on the side opposite to which the lower mandible crosses. For example, a bird with a lower mandible that crosses to its right secures seeds in the groove on the left side of its bill (see Fig. 4.3). It would be difficult if not impossible to husk a seed if it was secured on the other side, because the mandible crossing prevents the necessary alignment between the groove in the palate and the edge of the lower mandible that cracks the seed coat.

Initially, I didn't know if crossbills could recognize empty, unfertilized seeds without cracking open the seed coat. It seemed reasonable that crossbills could distinguish seeds that were empty from those with seed kernels by their weight. Indeed, it would be highly advantageous to discard empty seeds quickly, as often 10% or more of the seeds are empty (mostly because of self-pollination; see

FIGURE 1.7 Cassia Crossbills *Loxia sinesciuris* foraging on Rocky Mountain Lodgepole Pine *Pinus contorta* var. *latifolia* cones. (A) A male biting between the overlapping cone scales to form or enlarge a gap. Then (B) the lower mandible is spread to the side (abducted)—in this case to her right—to spread the scales farther apart to expose the seed, so that it can be lifted from the cone with the tongue (C). Note evidence (bulge) of large jaw muscles on the female's right side in B that provide power for jaw abduction. Crossbill jaw muscles are larger on the side to which the lower mandible crosses. Photographs taken in the South Hills, Idaho on 8 June 2004 (A and B) and 4 June 2016 (C).

Chapter 2, 'Seed ripening and shedding') and more time spent handling them is more time wasted by the bird. This wasted time adds up, because crossbills need to consume hundreds of to several thousand seeds a day, depending on the sizes of the seeds and the energy requirements of the crossbill (Benkman 2020).

By timing how long each seed was handled, and then gathering up all the discarded seed coats—counting both halves of the seed coat from husked seeds and slicing with a razor all the unopened seeds to determine if they were empty or not—I realized that foraging crossbills drop empty seeds whole (without cracking the coat) within a fraction of a second. Full seeds on the other hand required one and a half (or more) seconds to husk, depending on their size. Very occasionally a crossbill would watch a seed fall as if it had been dropped accidentally. It was only on those rare occasions—perhaps one in 1,000—that I would find an intact full seed. Thus, crossbills can assess whether a seed is full or not almost instantaneously and rarely make a mistake. Because few full seeds escape crossbills once they are in the bill (crossbills do not cache seeds), crossbills are truly efficient seed predators and thus strictly antagonistic in their interactions with conifers and their seeds.

When to leave a cone for another?

More than one seed is usually removed from a cone, but not necessarily all of them. That a few seeds are left has been noted by many observers who have inspected cones dropped by crossbills. Some of these cones might have been dropped prematurely by crossbills disturbed while feeding, perhaps by an aggressive crossbill or a passing predator. However, even undisturbed crossbills drop cones before all the seeds are removed. This has been described as 'wasteful'. In some cases, the occasional full seed left in the cone appears to have been simply missed (Tombre-Steen 1991).

Nonetheless, foraging theory explains how discarding a cone before all the seeds are consumed might well be an adaptive behavior by crossbills. For instance, individual seeds vary in their accessibility to crossbills depending on factors like the frequency of empty seeds or cone scale thickness, such that crossbills benefit from rapidly discarding cones that contain many empty seeds, or which have thicker scales that take more time for crossbills to spread apart. Similarly, it may be advantageous for crossbills to leave a cone when seed depletion means diminishing returns, and other cones are nearby and plentiful. Indeed, if one could quantify both the cumulative amount of energy gained while foraging on cones and the average travel time between cones, we could use the marginal value theorem (Charnov 1976) as a guide to test whether leaving a few seeds is indeed advantageous.

The marginal value theorem was developed for analyzing how long an animal should remain in a food patch before moving on to another. In the case of a crossbill, the cone can be considered a patch. As is commonly done, we can use

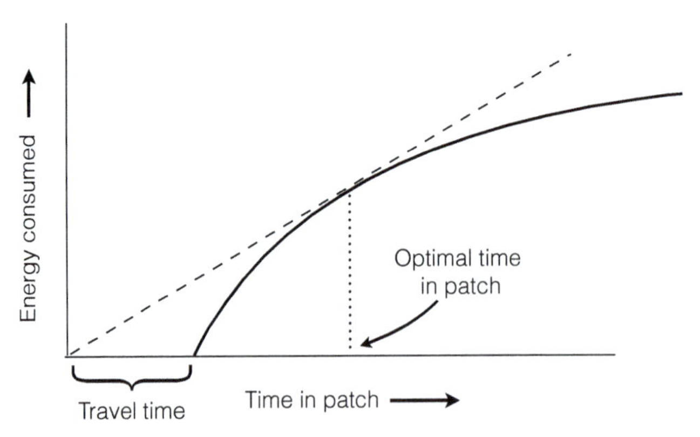

FIGURE 1.8 A graphical representation of the marginal value theorem used to predict the optimal amount of time to spend in a patch (e.g., a cone), based on both travel time and cumulative amount of energy eaten over time (the solid curve, termed a gain function) while foraging in a patch. The slope of the dashed line represents the overall rate of energy gain, and when the line is tangent to the gain function, the slope is maximized. Departing the patch earlier or later than the optimum results in a lower overall rate of energy gain.

a graphical model to illustrate optimal patch departing behavior (Fig. 1.8). The X-axis represents time, and the Y-axis represents cumulative energy gained (eaten). Starting at the origin, the animal travels to a food patch (requiring some amount of 'travel time'), and when the animal reaches a food patch it starts foraging. The curve representing the animal's energy gain rises as the animal consumes food, with the rate of increase usually depicted as decelerating over time (shown by how the energy consumption relationship is curved, rather than a straight line). This deceleration could occur for various logical reasons, including individuals retracing their steps as food in the patch is depleted, or initially consuming the most accessible, least wary or most rewarding prey. The optimal departure time from a patch can be solved by drawing a line from the origin that is tangent to the curve, as shown by the dashed line in Fig. 1.8. The slope of this line represents the energy gained per unit time by departing at the optimal moment, which is where the line is tangent to the curve. If the animal leaves earlier than the optimum, it spends proportionately more time traveling between patches when it could otherwise be consuming food at a relatively high rate, lowering its overall intake rate. Conversely, if the animal spends too long within a patch, the benefit of reducing the frequency of traveling between patches is outweighed by the diminishing rate of gains while in the patch.

I had the opportunity to test whether leaving seeds in cones might be adaptive when I spent the summer of 1987 color-banding White-winged Crossbills *Loxia l. leucoptera* in Ontario's boreal forest north of Lake Superior. White-wings had settled in an area where Tamarack had a large developing seed crop. Crossbills removed the small, closed cones from the branches (Fig. 1.9A), then extracted and husked seeds starting near the base of the cone, progressing toward the

FIGURE 1.9 (A) A young (<2 years old) White-winged Crossbill *Loxia l. leucoptera* perched in a Tamarack *Larix laricina* with relatively evenly spaced cones along branches. (B) Two open Tamarack cones illustrate the thin scales and the numbering of scales with full seeds from the base to the tip. The scales closest to the base and tip generally lack full seeds, but up to 11 scales had full seeds. A: photograph by Frank King in Newfoundland, Canada on 6 April 2024; B: photograph by William Friedman [https://commons.wikimedia.org/w/index.php?curid=90136776].

tip (distal end). Because the forest was dense and shaded, and the underlying moss moist, fallen cones did not dry out and open further. Thus, I was able to determine how many seeds were removed by crossbills, and from which scales, by examining recently dropped cones from under different trees. I was also able to measure (1) the average travel time between a crossbill dropping a cone and removing another, which was 2.1 seconds; (2) how long crossbills spent removing and husking each seed, which was 1.8 seconds; and (3) kernel masses under successive scales that I numbered from the base to the tip (Fig. 1.9B). Because of the fairly uniform distribution of the abundant cones on the branches (Fig. 1.9A) and the ease with which the small cones could be removed by crossbills, the average travel time I measured was likely representative for the crossbills present at that time. It was reasonable to use the average time to extract and husk a seed as representative of all cone scales because of both the uniformly thin scales throughout the cone (Fig. 1.9B) and the limited size variation in those seeds. Given these three sets of measurements, I was able to predict the optimal number of scales from which to remove seeds in a manner analogous to that described in Figure 1.8 and compare it to the number of scales from which crossbills actually removed seeds (Benkman 1989a).

What I found was a close, but not perfect, match between my theory-based predictions and the actual foraging behavior of these crossbills. My calculations indicated that the optimal behavior (that which maximizes kernel consumption per unit time) was for a bird to discard a cone after removing seeds up to and including the seventh scale. The overall intake rate decreased by leaving after the eighth scale rather than after the seventh because kernel mass decreased more consistently between scales seven and eight than between other successive scales. The real-life crossbills, on the other hand, most often dropped cones after removing seeds from the eighth scale—close to the optimum of seven, but not exactly.

Why didn't crossbills just count scales and leave after the seventh? One explanation is that crossbills can't count, or count that high. This is possible, but it doesn't explain why they left seeds at the distal end of the cone and departed most often after the eighth scale. Another explanation is related to variation in cone size. For example, it might be best for crossbills to continue foraging until they removed seeds from the eighth or ninth scale on the largest cones, and forage until the sixth scale on smaller cones. Then a more flexible strategy rather than strict scale counting might be favored. Consistent with this scenario, crossbills removed seeds from more scales as cone size increased. Crossbills might have adopted the simple rule of leaving the cone after they detected a decrease in kernel mass, which was often after foraging on seeds under the eighth scale. Assuming the ability of crossbills to detect differences, even slight differences, in kernel size is not unreasonable. Such a simple rule would serve them well. I estimated their overall intake rate to be 99% of the potential intake if they had left one scale earlier (Benkman 1989a). Therefore, leaving a few seeds is not wasteful from a crossbill's point of view!

Which way to cross

I am often asked whether the lower mandible crosses to the left in some crossbills and to the right in others, and if so whether the frequencies are equal. The answer is 'yes' and 'sometimes'. The lower mandible crosses to the right in some individuals and to the left in others (e.g., individuals in Figs 1.7B,C and individuals in Figs 1.3 and 1.4, respectively), and the frequencies of each are approximately equal in at least some taxa of crossbills. That's easy to see and summarize. The answer to the follow-up question of 'why?' requires a bit more thought and explanation. And before I address the 'why', I should note that the direction to which the lower mandible crosses appears set in the nestling stage, if not earlier in embryonic development based on observations that Harrison 'Bud' Tordoff shared with me years ago. When Bud placed his fingers on the sides of the bills of nestling Red Crossbills from his captive breeding colony, the nestlings always abducted their lower mandible to the side to which it later crossed, which became apparent several weeks after they fledged.

Now back to the 'why' question concerning variation in the directions to which the lower mandibles cross. First, the mandible crossing direction need not vary among individuals. The Wrybill *Anarhynchus frontalis*, a New Zealand shorebird, has a bill whose tip curves to the right. Always. This is termed directional asymmetry. Directional asymmetry occurs in a diversity of organisms, from claw asymmetries in swimming crabs, to chirality in snails (the direction their shell spirals) and the asymmetry of our hearts. In contrast, the 1:1 ratio of left to right crossings in some crossbills is termed antisymmetry. It occurs in a diversity of organisms too, including asymmetries in fiddler crab claws and flower structures (Palmer 2016). The direction of asymmetry in antisymmetric species is rarely inherited (Palmer 2004), and this appears to be the case in Red Crossbills based on parent–offspring relationships (Edelaar et al. 2005). Presumably there is a mechanism that randomizes the direction of crossing during development so that there is no relationship between the mandible crossing direction of parents and offspring, and thus there is no 'heritability'. The result is an equal frequency of left and right crossings in the population. Given that both directional asymmetry and antisymmetry are common and widespread in animals, is there a reason why crossbills are antisymmetric rather than directionally asymmetric as in Wrybills? I think the answer has to do with the potential for food competition arising from the stereotypic foraging behavior of crossbills.

The stereotypic way that crossbills forage on closed or mostly closed cones results from the difference in the structure of their upper and lower mandibles (Fig. 1.3), which enhances their ability to spread apart scales (Benkman 1987b). A consequence of this consistency in orientation is that from a given perch, the most accessible seeds for a foraging crossbill are those on the side of the cone opposite to which the lower mandible crosses (Fig. 1.10). For example, the most accessible seeds to a crossbill whose lower mandible crosses to its right are those on the left side of the cone to the approaching crossbill (as in the right side of Fig. 1.10). If perch sites near the cone are limited and a bird cannot easily access all sides of the cone from different perches, the readily accessible seeds might even

FIGURE 1.10 Mirror images of a Cassia Crossbill *Loxia sinesciuris* foraging on a Rocky Mountain Lodgepole Pine *Pinus contorta* var. *latifolia* cone (as in Fig. 1.7B). When crossbills forage for seeds in closed or reclosed cones, they forage on the side of the cone opposite the side to which their lower mandible crosses. Thus, the crossbill on the left side of this image forages on the right side of the cone (as viewed by the crossbill) because its lower mandible crosses to its left, and vice versa for the crossbill on the right side of the image.

be non-overlapping between the two crossing types. An analogous situation occurs in African scale-eating cichlid fishes whose mouths either curve to the right or left, and who approach either from the left or right of the fish they are parasitizing, respectively, from which they bite off scales (Hori 1993).

Crossbills in at least some populations often encounter cones attached to branches that have been partially depleted by other flock members, or by other flocks in earlier visits. In this situation, a crossbill with the uncommon crossing type would be at an advantage, as it would be more likely to encounter cones foraged on by a crossbill with the opposite crossing to its own. Individuals with the uncommon crossing type should thereby experience a feeding benefit, because they are more likely to encounter cones that still contain the seeds that are most accessible to them. Crossbills should then experience what is termed 'negative frequency-dependent natural selection', where selection experienced by an individual depends on the frequency of the different traits in the population, with the uncommon type having the advantage (Fig. 1.11). However, if mandible crossing direction is not heritable, such selection will not result in evolution. Evolution in response to selection requires that the trait is heritable, and, as mentioned above, mandible crossing direction might not be heritable. Nevertheless, negative frequency-dependent natural selection would drive the ratio of crossing types toward 1:1 within each generation and cause selection

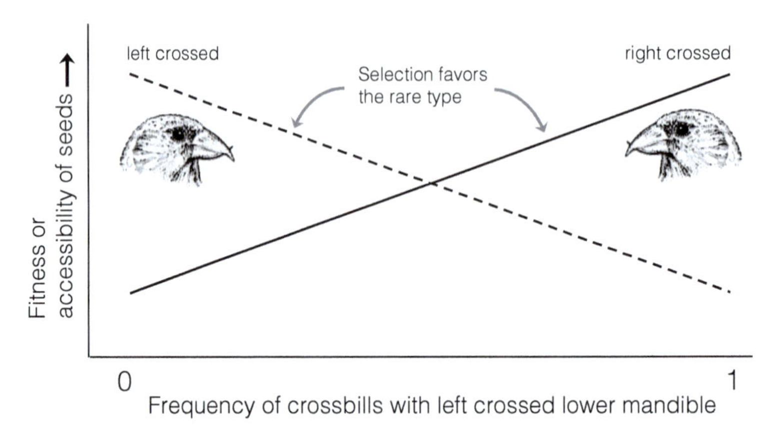

FIGURE 1.11 Negative frequency-dependent selection occurs when the fitness of individuals decreases as the class to which a given individual belongs increases in frequency relative to other classes of organisms. In the case of crossbills, the fitness (i.e., relative accessibility of seeds in cones) of individuals whose lower mandibles cross to the left decreases as the frequency of such individuals increases in the population (dashed line; it is linear for simplicity). Conversely, the fitness of individuals whose lower mandibles cross to the right increases with the frequency of individuals whose lower mandibles cross to the left (solid line). As the frequency of crossing type drifts from one to one (where the two lines intersect), the rarer type will have higher fitness, resulting in an increase in its occurrence with the frequency stabilizing around 0.5 (a 1:1 ratio).

against mutations that bias the direction of crossing, such that the two crossing types remain equally common.

Although negative frequency-dependent natural selection is especially prevalent in the context of resource competition (Gómez-Llano et al. 2024) and struck me as logical for foraging crossbills, it is best to test this hypothesis. Such a test would be formidable in the wild, but it is far more feasible with captive crossbills. The experimental design is simple (Benkman 1996): time how long a crossbill requires to remove and consume a set number of seeds from a cone, and then do the same on the same cone for a second crossbill of either the same or opposite crossing direction. Then repeat this for multiple cones and multiple birds to increase one's confidence in the results.

If negative frequency-dependent natural selection while foraging accounts for the 1:1 ratio of mandible crossings, then the prediction is that intake rates would be worse for the second crossbill when its mandible crossing is the same as the first crossbill, and better for the second crossbills that have bill-crossing directions opposite that of the first bird. Fortunately, we had ideal cones for the experiment. The previous fall, Chris Milensky and I had conducted fieldwork in the Cypress Hills in southeastern Alberta just north of the US border. Chris, who is now the Collection Manager for birds at the National Museum of Natural History in Washington, DC, was a recent graduate of New Mexico State University where I was a faculty member. I had hired Chris to help with field studies and cone measurements, and with the crossbill experiment.

We had gathered hundreds of serotinous Rocky Mountain Lodgepole Pine *Pinus contorta* var. *latifolia* cones. Serotiny is a widespread and common trait among Rocky Mountain Lodgepole Pine, where the cones often remain closed for many years—even over decades—until heat, generally from a fire, melts the resinous bonds that effectively glue them. This allows the cones to open and shed their seeds into a freshly burnt landscape with promising opportunities for seedling survival and growth. Seeds in the closed cones are inaccessible to crossbills. Only after the scales have begun to separate, even ever so slightly, can crossbills gain access to the seeds. In our study, instead of sparking a forest fire, we boiled the cones in water for a minute or two to open the scales. Then we placed the cones in plastic bags to keep the scales moist and to reclose the scales, securing the seeds within. The cones were ideal for the experiment because crossbills could regularly and dependably remove seeds from between the uniformly reclosed scales. Chris then drilled a small hole in the base of the cones and secured them one at a time to a screw at the end of a wooden dowel so that there was only one perch on one side of the cone, and therefore all birds had to access it from the same perch orientation. He then timed each hungry crossbill removing and consuming 10 seeds, which represented about a quarter of the total seeds in the cone. Next Chris turned off the aviary lights and captured the first crossbill. He then turned on the lights and released a second crossbill into the aviary and ran his stopwatch until it, too, had consumed 10 seeds.

The bill orientations of the two birds made all the difference to the crossbill that came second in the queue, as the intake rate of that second crossbill was lower only when it followed a crossbill with its own mandible crossing direction. There was no depression at all in intake rate when following an individual of the opposite crossing. This provides a mechanism that would favor the rarer mandible crossing type and account for why the pine-feeding crossbills in the New World have approximately equal frequencies of left to right crossings. If a mutation arises that would bias the mandible crossing direction to one side, then individuals carrying that mutation would be at a feeding disadvantage resulting in selection against individuals carrying that mutation. This will act to maintain an equal frequency of left and right crossings over time. For example, we find a 1:1 ratio in Cassia Crossbills *Loxia sinesciuris*, which feed almost entirely on seeds in serotinous Rocky Mountain Lodgepole Pine cones that are never removed from the branch and are quite similar to those that we used in the foraging experiments (Fig. 1.7). Of the 3,261 Cassia Crossbills that we have banded over the years, the lower mandible crossed to the left as often as to the right (1,644 or 50.4% and 1,617 or 49.6%, respectively; Chi-square = 0.224, $P = 0.64$).

Crossbills, however, are unlikely to encounter partially foraged cones where crossbills remove closed cones from branches, with the cone falling to the ground afterwards. Cone removal is most common when crossbills forage on small cones, and is especially prevalent in Europe. Indeed, a sample of 1,627 crossbills captured at one location in the Pyrenees in Spain, where foraging crossbills nearly always remove cones, differed from a 1:1 ratio of mandible crossings (879 [54%] to the left, 748 [46%] to the right, Chi-square = 10.55, $P = 0.001$; Senar et al. 1995). Why do foraging crossbills usually remove cones in Europe? It might be related to the absence of native pine squirrels *Tamiasciurus* spp. on that continent. Although there are tree squirrels in Europe, namely European Red Squirrels *Sciurus vulgaris*, they do not harvest the vast numbers of cones that pine squirrels harvest in North America, and thus do not exert as strong an effect of natural selection on the cones, including adaptations that drive them to have stronger peduncles securing cones to branches. Consequently, cones might be easier to remove in Europe, explaining why crossbills more regularly remove them there than in North America. The general impact of pine squirrels on cones, and thereby on crossbills, is considerable and forms the focus of Chapter 5.

A further contrast to New World pine-feeding crossbills like Cassia Crossbills is the White-winged Crossbill described earlier (Fig. 1.9A), because it removes the small spruce and Tamarack cones that it relies upon (or twists the attached cones so that all sides are equally accessible). For white-wings, there is no benefit to, nor should there be natural selection favoring, an equal ratio of left to right mandible crossings. Not surprisingly, the ratio of left to right mandible crossings differs from 1:1 in a sample of birds I have examined: 218 (27.8%) crossed to the left and 566 (72.2%) crossed to the right (Chi-square = 154.5, $P < 0.0001$; Benkman

1996). Presumably mutations that bias the direction of mandible crossing away from random are not selected against in white-wings. Intriguingly, the Hispaniolan Crossbill, which evolved from a white-wing-like crossbill ancestor (Fig. 1.5), relies on seeds in pine cones that it does not remove, and its mandible crossing ratio again approximates 1:1 (59 [49.2%] to the left, 61 [50.8%] to the right, Chi-square = 0.033, $P = 0.86$).[2]

A cost of specialization

The crossed mandibles and associated musculature have enabled crossbills to exploit otherwise difficult to access seeds, and in turn have led to a radiation of different crossbill taxa (Fig. 1.5). Such specialization, however, potentially has its downsides, such as when seed crops periodically fail over large regions (Chapter 2) or when the abundances of conifers decline from drought, beetle-kill, fire or logging (Chapter 9). But to understand some of the downsides or costs to crossbills, let's step back and consider their evolution.

Crossbills evolved from within a clade (a related group that all share a common ancestor) of finches whose extant species are mostly about half the size of a crossbill. For instance, crossbills are most closely related to Redpolls (Fig. 1.2), which average about 14 grams, while most crossbill taxa average between 27 and 40 grams. The crossbill-Redpoll clade is sister to linnets *Linaria* spp. (Fig. 1.2), whose body masses average between 13 and 20 grams. Thus, it seems reasonable to infer that crossbills evolved from a relatively small-bodied ancestor.

In seed-eating birds, larger body size and larger bills are generally associated with the ability to exert stronger biting forces (van der Meij and Bout 2008) and with eating larger seeds (Newton 1967; Grant 1986). Crossbills, however, do not eat particularly large or hard seeds; most seeds that crossbills eat weigh less than 20 mg. Relative to their body size—a good metric reflecting the biting forces they can exert when cracking seeds (van der Meij and Bout 2008)—crossbills consume small seeds when compared to other cardueline finches that primarily feed on non-conifer seeds (Newton 1967). Crossbills have apparently evolved a larger size because larger bills, skulls and associated musculature are needed to provide the forces necessary to pry open woody cones, rather than to husk the seeds inside them.

To determine whether there is a cost of specialization for crossbills, I compared the feeding efficiencies of crossbills to other cardueline finches when feeding on non-conifer seeds. What I mean by feeding efficiency is an individual's effectiveness at meeting its daily energy requirements, because I needed to consider

2 This includes data from Benkman 1996, plus crossing direction gleaned from photographs available from the Macaulay Library at the Cornell Laboratory of Ornithology (accessed on 24 January 2024), from which I could distinguish a total of 93 different individuals.

the relatively large body sizes and hence elevated energy demands of crossbills. The most efficient individual would spend the least amount of time foraging to meet their energy demands. Ian Newton recognized the value of this approach when he noted the limitations of using just measures of seed handling ability for comparisons between species that varied in body size. Ian wrote: 'A more pertinent experiment would be to measure the daily food requirements of different finches and see which size of seed would yield these most quickly for each species' (1967: 71). So first, I quantified seed handling ability (intake rate) of White-winged and Red Crossbills (27 and 33 gm, respectively), and three other cardueline finch species spanning the body size range of crossbills—American Goldfinch *Spinus tristis* (13 gm), House Finch (21 gm) and Evening Grosbeak (55 gm) (Fig. 1.2)—feeding on a range of seed sizes (2.3 to 117.5 mg) from small thistle to large sunflower seeds (Benkman 1988). I used thistle and sunflower seeds because all the finches readily fed on them, I could get a wide range of seed sizes and they are composite seeds (Asteraceae), which are an important natural food for many cardueline finches (Martin et al. 1951; Taber 1968; Newton 1967, 1972). Then I used allometric relationships from the literature to estimate daily energy requirements (Walsberg 1983) and solved for the necessary encounter rate of seeds (effectively seed abundance) to meet daily energy demands for each combination of finch species and seed size (Benkman 1988).

Although crossbills handled the smaller seeds as well as the small goldfinch and House Finch, crossbills were much less efficient at meeting their energy demands because of their large body masses, and hence higher energy requirements. Crossbills were less efficient on larger seeds because they spilled so much kernel when breaking the large kernels into smaller pieces to swallow. This spillage of kernel is related to their specialization for feeding on seeds in conifer cones. Many cardueline finches have palates that are recessed relative to the cutting edge on the sides of the upper mandible (maxillary tomium; House Finch in Fig. 1.12; Ziswiler 1965). A recessed palate results in a larger cavity or buccal volume in which seeds can be mandibulated and broken prior to swallowing, without pieces of kernel spilling out of the mouth. The palates of crossbills are less recessed, presumably to aid in handling small seeds, and perhaps to provide greater strength for a given mandible depth. But a smaller mouth cavity results in crossbills suffering an increasing amount of kernel loss as seed size increases (Benkman 1988). Further contributing to a small mouth cavity are their slender mandibles (Fig. 1.12), which are an adaptation for sliding with less resistance between cone scales and limiting the width of the gap crossbills need to create to reach seeds.

Overall, I estimated that a crossbill requires an abundance of non-conifer (composite) seeds that is about two to three times greater than that required by other cardueline finches (Benkman 1988). Because these other finches likely reduce the abundance of non-conifer seeds over winter through consumption, crossbills are unlikely to fare well if they need to rely on such seeds for long. This is a cost to specialization for crossbills because they cannot rely on non-conifer seeds in

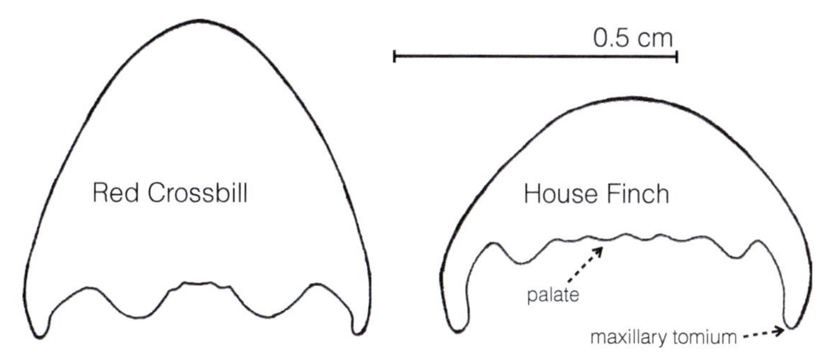

FIGURE 1.12 Cross-sections of the upper mandibles (at the anterior edge of the nares) of a Red Crossbill *Loxia curvirostra* (33 gm body mass) and House Finch *Haemorhous mexicanus* (21 gm) illustrate the former's relatively narrow mandible for its body size and lowered horny palate in respect to the outer edges of the mandible (the maxillary tomia). The wider mandible and recessed palate of the House Finch provides a larger cavity in which to break large seeds into smaller, swallowable pieces without spilling them. From Benkman (1988).

winters when the conifer seed crop fails. And it puts a premium on being able to disperse to locate large conifer seed crops, on which crossbills have a feeding advantage compared to other birds, as described earlier. The challenges of locating large conifer seed crops are a focus of the next chapter.

Chapter 2

Challenges and Opportunities of Relying on Cone-seed

> They [crossbills] are abundant in a locality one season, rare the next, then entirely absent for a season. They are most unreliable.
>
> (W. Taber 1968: 517)

It is one thing to be able to extract seeds from cones. It is another to be able to find conifer seeds year in and year out. As the last seeds are depleted and shed from one seed crop, crossbills need to find new seeds somewhere—either in an alternative conifer species, in a new area where seeds are still held, or (if in summer) in a newly developing seed crop. In some years, finding cone-seed can be especially challenging. Large seed crops for particular conifer species are often followed by seed failures that occur synchronously over vast areas (Koenig and Knops 2000, 2001; LaMontagne et al. 2020). This explains why most crossbill populations are nomadic, with individual crossbills sometimes wandering far in search of a seed crop. Indeed, crossbills may fly up to 1,600 miles (2,600 km) or more between their natal and breeding sites and between breeding sites (Newton 2006). I began to appreciate this challenge of finding successive seed crops during my second year studying crossbills, when I couldn't find a large seed crop, and nor could I find more than a few crossbills within hundreds of miles of my campus. As Taber (1968) noted, they can be 'most unreliable'!

Many conifers produce a large seed crop only every two to six years (Burns and Honkala 1990). When conifers do produce large seed crops in successive years, seed predators, especially insects, dramatically increase in abundance while feasting on the first seed crop. The now large seed predator population leaves few surviving seeds in the following year's crop (Keen 1958). Because of this numerical response by seed predators, they exert natural selection that favors the production of variably sized seed crops from year to year, a widespread phenomenon in plants that is termed 'masting' (Box 2.1). If small or no seed crops occur prior to a large seed crop, then seed predator populations will be small and become satiated as the large seed crop matures, allowing plenty of seeds to escape predation, disperse and recruit (e.g., Linhart et al. 2014).

Box 2.1: Annual variation in seed crops

Multiple factors, including enhanced pollination resulting from synchrony in the production of pollen- and seed-cones, might favor variation in the size of seed crops from year to year (Smith et al. 1990). Nonetheless, I suspect that lessening seed predation is a widespread and primary factor favoring seed crop fluctuations in many cone-bearing conifers (e.g., Fletcher et al. 2010 for White Spruce *Picea glauca*) as has been documented for many plants (Zwolak et al. 2022). I am particularly persuaded by geographic variation in seed crop fluctuations in Rocky Mountain Lodgepole Pine *Pinus contorta* var. *latifolia*. In Colorado and Wyoming, where the primary predispersal seed predator of Lodgepole Pine, the American Red Squirrel *Tamiasciurus hudsonicus*, is common, the coefficient of variation in annual seed crops is in the range of 70% to 90%. Coefficients of variation are standardized measures of variation (the standard deviation divided by the mean, times 100), with values for annual variation in conifer seed crops tending to range from a low of about 70% to a high of 150% or even 200% (Kelly and Sork 2002). Remarkably, in small mountain ranges to both the west and east of the Rocky Mountains where American Red Squirrels are absent, the coefficient of variation for Lodgepole Pine is exceedingly small at 5% to 7%, reflecting that the variation between years is negligible (Benkman et al. 2003, 2013). These values are about a third of the smallest recorded value for plants tabulated in an extensive literature review by Kelly and Sork (2002).

Even though fluctuating seed crops are an effective defense against many seed predators (Zwolak et al. 2022), these predators have evolved corresponding counter-adaptations. Some insects remain in diapause for multiple years, thus bypassing the poor seed crops immediately following a large seed crop while waiting for the good times to return with the next large seed crop (Turgeon et al. 1994). Crossbills deal with this variability through their powerful flight and nomadism, which allow them to track spatiotemporal variation in seed crops far more effectively than (for example) relatively sedentary tree squirrels (Summers 2011). Nevertheless, nomadism does not ensure that crossbills can always find large seed crops, especially during large-scale cone failures, which undoubtedly cause much mortality in crossbills. It is difficult to make accurate population counts of these non-territorial, nomadic crossbills and assess how much mortality occurs following cone failures, but my impression through following White-winged Crossbills *Loxia l. leucoptera* in the 1980s is that it takes several years before populations return to pre-eruption numbers (i.e., the number of crossbills just prior to the cone crop failure). The one study that has been able to estimate population declines following eruptions was on Red-breasted Nuthatches *Sitta canadensis* erupting in the fall out of the boreal forests in eastern North America (Dunn 2019). The author, Erica Dunn, estimated that nuthatch breeding populations declined by up to 50% following the largest eruptions. I suspect similar if not greater declines occur in crossbills, because crossbills are even more reliant on conifer seeds than are nuthatches and thus are likely to experience higher mortality during large-scale cone failures.

Nomadism also poses challenges for study. It is not surprising that most studies of crossbills are unlike the classic multi-year ornithological studies on color-banded populations. Most field studies of crossbills have instead been opportunistic. Nevertheless, we have a unique window into crossbill natural history because of both our ability to measure feeding intake rates and the tractable and close link between crossbills and conifer seed crops.

Seed ripening and shedding

> In any study of evolutionary ecology, food relations appear as one of the most important aspects of the system of animate nature. There is quite obviously much more to living communities than the raw dictum 'eat or be eaten,' but in order to understand the higher intricacies of any ecological system, it is most easy to start from this crudely simple point of view.
>
> (G.E. Hutchinson 1959: 147)

To understand Hutchinson's 'food relations' for crossbills, it is best to start with the phenologies of seed and cone ripening, and seed shedding. Crossbills occur mostly in temperate regions in the Northern Hemisphere (Newton 1972). Here, seed cones are initiated in late summer, but with little development until the following spring when the buds break and wind pollination occurs (Fig. 2.1A). In larch *Larix* spp., spruce *Picea* spp., Douglas-fir *Pseudotsuga menziesii* and hemlock *Tsuga* spp., fertilization occurs soon after pollination, and the cones mature over the ensuing summer. In pine *Pinus* spp., fertilization and most development is delayed until the second and in a few species the third spring and summer (Fig. 2.1B). For both groups of conifers, seed coats develop after ovules are fertilized. The kernel—female gametophyte and embryo—develops if pollen fertilizing the ovule is from another tree (outcrossing or cross-pollination), but not so often when self-pollinated (cone-bearing conifers in the pine family, Pinaceae, which includes pine, spruce, larch, fir, hemlock and Douglas-fir, are monecious, with individuals producing both pollen cones and seed cones), which then results in empty seeds that have a seed coat but no nutritious seed within. In large stands, most ovules receive pollen from other trees, but that is not the case for isolated trees, which may produce mostly empty seeds (Smith et al. 1988; Robledo-Arnuncio et al. 2004). In general, pollen production, outcrossing and thus the number of full seeds per cone increase with the size of the cone crop (Smith and Balda 1979). By July or early August, the seeds of many conifers are sufficiently developed for the crossbills to profitably forage on them (Fig. 2.2).

FIGURE 2.1 On the left (A) are conelets of Rocky Mountain Lodgepole Pine *Pinus contorta* var. *latifolia* in July soon after pollination but a year before they develop rapidly and mature. On the right (B) are immature cones of Rocky Mountain Ponderosa Pine *Pinus ponderosa* var. *scopulorum* in April that were pollinated the previous late spring–summer. They have developed slightly since pollination but most development occurs in the next four months; the open cone behind the needles in the lower left is a cone that matured a year earlier. Three or more conelets on many of the branch tips on most trees indicate a large future seed crop that is likely to attract crossbills.

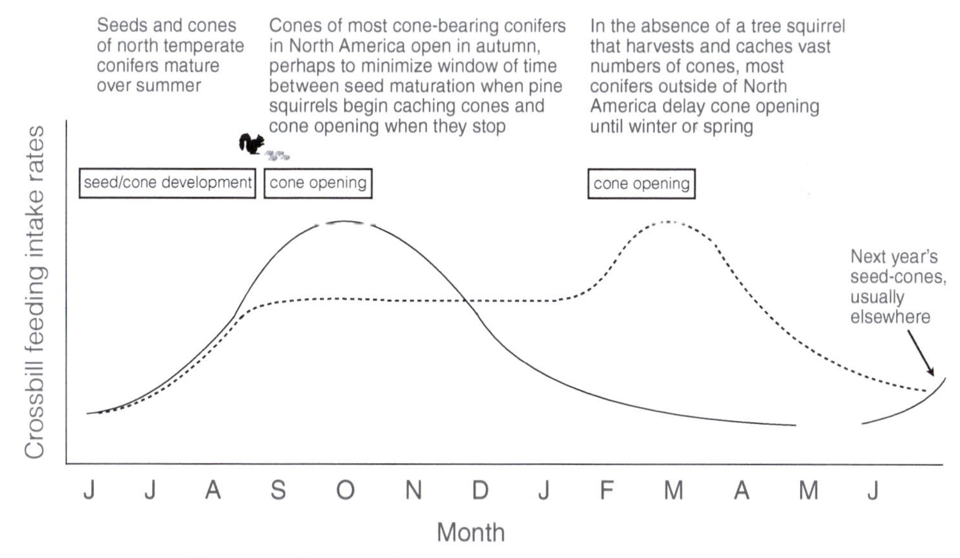

FIGURE 2.2 A schematic of seasonal variation in crossbill feeding intake rates. Intake rates increase as seeds and cones mature over summer, then peak when cones open and seeds are both plentiful and accessible. Intake rates then decline as seeds are shed and depleted from open cones. Separate curves are shown for areas with and without pine squirrels *Tamiasciurus* spp. (solid and dashed lines, respectively), because the cones of most conifers within the range of pine squirrels (North America) open in autumn, whereas those elsewhere generally open in winter or spring.

As the cones reach their final sizes, the seeds mature and within several weeks the cones of most conifers in North America (north of Mexico) dry and the scales begin separating, releasing seeds. This is when crossbill feeding intake rates peak, because seeds are full-sized and readily accessible under most scales (Fig. 2.2). Earlier, seeds are plentiful but less accessible and, depending on how much earlier, smaller in size and less nutritious. As seeds are released, they become increasingly difficult to locate in the cones and intake rates decline for foraging birds. Crossbills generally don't feed on loose seeds that have fallen to the ground. Exceptions include Red Crossbills *Loxia curvirostra* feeding on the large seeds of some western North American pines including Single-leaf Pinyon *Pinus monophyla* (409 mg seeds; Payne 1972) and Ponderosa Pine *P. ponderosa* (37 mg seeds), which individually provide large rewards and which are easier to locate among leaf litter (Fig. 2.3); otherwise, crossbills generally eat conifer seeds that weigh under 20 mg.

The rate at which seeds are released varies among conifer species, and among locations and years. An important variable influencing the release of seeds within a conifer species is the occurrence of warm, dry and windy conditions, resulting in early and rapid seed release, or cool, moist conditions, resulting in delayed seed release (Burns and Honkala 1990; Mezquida et al. 2018). This variation is not shown in Figure 2.2 but can be envisioned by shifting the rate at which intake

FIGURE 2.3 Male Red Crossbill *Loxia curvirostra* husking a Ponderosa Pine *Pinus ponderosa* seed that he had picked out of the leaf (pine needle) litter. Photograph taken by Kadi Franson in Bryce Canyon National Park, Utah on 6 May 2020. I saw the same foraging behavior by a flock of Red Crossbills in Rocky Mountain National Park, Colorado in the same month and year; this is also when few seeds remain in the cones.

rates decline after cones open: a more rapid decline if seeds are shed quickly, a more gradual decline if seeds are shed slowly (see Fig. 9.12 for an example). The implications of climate change are probably obvious and will be explored in Chapter 9. Seed releases also seem delayed for thick-scaled cones, perhaps because their scales open less readily during warm, dry conditions than those of thin-scaled cones, but also because their cones tend to open later (Smith 1970).

In Eurasia and Mexico, seeds and cones mature in summer as in North America, but, instead of opening in the fall, the cones generally remain closed until December or until the following March or April (Fig. 2.2). I believe the occurrence of pine squirrels *Tamiasciurus* spp. in North America (north of Mexico) and their extensive harvesting and caching of mature, closed cones accounts for the difference in ripening phenology between regions. The brief time window between seed maturation, when pine squirrels start intently harvesting and caching cones, and cone opening, when pine squirrels cease harvesting cones (Fig. 2.2; Smith 1968, 1970), saturates their caching abilities and limits their main cone harvesting activities to several weeks; pine squirrels are strictly a seed predator and not a seed disperser (Smith 1970). Tree squirrels in the genus *Sciurus*, but not pine squirrels, occur in Eurasia, Central America and Mexico (pine squirrels occur in a small area in northern Baja California, but do not cache large numbers of cones as they do elsewhere; Koprowski et al. 2006). However, *Sciurus* do not harvest or cache anything close to the vast numbers of cones cached by pine squirrels; pine squirrels typically store most of their cones in a central location or midden within their territory (Smith 1968), termed larderhoarding, whereas *Sciurus* scatter hoard cones if they cache them. Thus, selection for an abbreviated period between seed maturation and seed shedding is much weaker in areas without pine squirrels than in areas with pine squirrels. Presumably in the absence of pine squirrels, selection that favors delayed seed dispersal (e.g., via the numerous seed predators that feed on seed in either open cones or those that have fallen to the ground, or via damaging mold in the soil) counters natural selection for early seed release as a result of predispersal seed predation by crossbills and tree squirrels other than *Tamiasciurus*.

Crossbill responses

In response to the seasonal development and annual fluctuations in seed crops, crossbills move in search of developing seed crops starting in late April and May and continue searching until a large seed crop is located (Newton 1972; Benkman 1987a, Cornelius and Hahn 2012). This pattern is so regular that nomadic crossbills preemptively accumulate fat each spring (Cornelius and Hahn 2012), providing energy to fuel long-distance movements across potentially inhospitable terrain, as in many seasonally migratory birds.

How far crossbills move depends on the distances between successive large seed crops. These distances vary from year to year and among regions and taxa of crossbills, because of variation in the geographic extent of seed crops, and variation in the intervals between suitably large seed crops. The spatial extent to which conifer seed crops are synchronous depends in part on the uniformity of climate across the region, because similar weather can act to synchronize seed production. In particular, an increase in temperatures from one summer to the next appears to initiate the production of large seed crops, with the constraint that large seed crops generally do not occur in successive years. For instance, large White Spruce seed crops occur in years that follow a July with temperatures higher than in the previous July (LaMontagne et al. 2020). Because weather in the relatively flat northern boreal forests can be synchronous across vast areas, the geographic extent of large seed crops and conversely seed failures can be widespread, occurring across 600 miles (1,000 km) or more (Koenig and Knops 2000; Koenig 2001; LaMontagne et al. 2020). This forces crossbills and other seed-eating birds to move long distances from one year to the next to find large seed crops. In contrast, where local variation in weather is much greater, such as in the Rocky Mountains, the spatial extent of large and small seed crops is often smaller (Mooney et al. 2010), so that annual movements by crossbills presumably average shorter (Bock and Lepthien 1976).

Long movements were evident when I conducted research for my dissertation (Benkman 1987a). White-winged Crossbills, which are denizens of the northern boreal forest extending from Alaska across Canada to the Maritime provinces, were common in the winter of 1982–3 in southern Ontario and east to the Laurentian Mountains of Quebec. But in the following year, white-wings were scarce in most of eastern North America. They were common that year only in Alaska and the Yukon, where there was a massive White Spruce seed crop. Much breeding occurred in those areas, driving an overall increase in their population. The following late May and early June (1984) white-wings departed Alaska and the Yukon, where few cones were developing. Because of their large numbers, white-wings could be tracked from reports in *American Birds*. Some moved south down the Rocky Mountains, but most moved eastward. After one to two months, many arrived in southern Ontario east to the Maritime Provinces and New England, where there was a large developing White Spruce seed crop. Many of these individuals arriving in the Maritimes likely originated in Alaska, which represents a flight of at least 2,800 miles (4,500 km)! This pattern of large White Spruce seed crops in Alaska and the Yukon in one year followed by a large seed crop in eastern Canada the next year and vice versa appears to occur regularly (LaMontagne et al. 2020).

How did white-wings locate large seed crops in eastern Canada and New England? It could have been pretty simple: the birds might have followed the belt of boreal forest that extends across Canada until they found large seed crops. Movements at this time of year coincide with longer, warmer days and an increasing abundance of insects that the crossbills can eat to supplement the relatively few seeds remaining in cones. Black Spruce *Picea mariana* holds seeds in its partially closed cones year-round (its cones are semi-serotinous with some seeds held for nearly 20 years; Chai and Hansen 1952)—seeds that in combination with plentiful and accessible insects would be sufficient to fuel long flights. Such movements east (and west) across the boreal forest are probably a regular occurrence in white-wings, like the analogous 'pendulum flights' described by Svärdson (1957: 330) for Red Crossbills in the northern forests of Eurasia, and for other boreal seed-eating finches in North America (Strong et al. 2015; Hannah et al. 2020).

It seems logical that crossbills would attempt to remain in conifer forests while moving in search of seed crops, but the directions crossbills set off in and how these might change are unknown. Given that I can anticipate where future seed crops of pine will occur a year before based on observations of recently pollinated conelets (Fig. 2.1), I would be surprised if crossbills don't use similar information to help guide their travels. If individuals have not traveled through areas portending large future cone crops, then they might follow other individuals, some of whom might have flown through such areas. This might account for crossbills amassing into unusually large flocks before leaving a region (e.g., Marquiss and Rae 1994). This would represent a form of information sharing thought to be important for migration in many species (Bolger et al. 2007). I suspect that substantial advances in our understanding of crossbill movement strategies are not far off. The limiting aspect might be the lack of good information on spatiotemporal variation in seed crops over range-wide scales. If we had such information, we could model different search strategies that can be tested against actual crossbill movements. Such data are inevitable with advances in remote sensing and animal tracking technology.

In North America, a second period of crossbill movement occurs in late October and November (Benkman 1987a). Autumn movements are less regular than in late spring because they depend on the size of the seed crop and the extent of seed shedding that has occurred. When seed crops are large, a high proportion of seeds are full, and crossbills are more likely to be supported through winter. When seed crops are small, seed will likely be insufficient for surviving the winter unless only a few crossbills remain. Movements are most pronounced when seed crops are small and crossbills are numerous, and sometimes lead to crossbills irrupting into areas outside their usual ranges. Indeed, the size of irruptions increases with the size of the crossbill population (which increases with the size of the previous year's seed crop) and decreases with the size of the current seed crop (Koenig and Knops 2001; Summers et al. 2024).

By measuring intake rates of white-wings throughout the year, I found that they left areas in late October–November when their intake rates declined below 0.4 mg of kernel per second. But they remained when their intake rates were higher (Benkman 1987a). Although their intake rates were more than sufficient for survival during autumn (I estimated they required an intake rate of 0.2 mg/second during January, when energy demand per unit time available for foraging is greatest), the risk of being unable to survive the winter with further seed fall and seed depletion presumably favored white-wings that were unwilling to take such a risk. On the

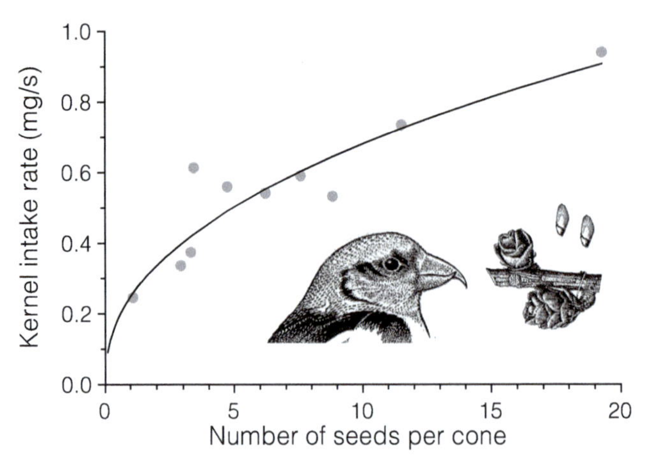

FIGURE 2.4 White-winged Crossbill *Loxia l. leucoptera* kernel intake rates increase with increases in the number of seeds in Tamarack *Larix laricina* cones. Each data point represents the mean rate and mean number of seeds per cone at a given location over one to several days. The curve is the best-fit logarithmic equation ($r^2 = 0.83$). From Benkman (2024).

other hand, higher intake rates implied more seeds per cone (Fig. 2.4), and thus a greater likelihood that the seed supply would remain sufficient to support crossbills over winter.

Although intake rates generally peak as cones open, the timing and magnitude of peak intake rates and the subsequent rate of decline vary among conifers (Benkman 1987a). Not surprisingly, crossbills track such variation so as to maximize their intake rates, much as I found for the cone 'departure' decisions by white-wings discussed in Chapter 1, 'When to leave a cone for another?'. For example, the timing of cone opening differs among the three species of spruce in eastern North America with corresponding effects on white-wing intake rates. White Spruce has the thinnest scales and opens first. Several weeks later Red Spruce *Picea rubens* cones begin to open, before finally some of the Black Spruce cones open with many not opening until the following spring.

When there were large seed crops of all three spruce, such as I found in New Brunswick, Canada in 1988–9, crossbills foraged sequentially on these species. Before any of the cones opened, white-wings foraged exclusively on White Spruce, which has the thinnest scales and the most accessible seeds when cones are closed (see Fig. 3.1). Only after these cones opened and with some seed depletion and intake rate decline did white-wings switch to foraging on the opening Red Spruce cones (Fig. 2.5). Red Spruce, with thicker cone scales than White Spruce, tends to shed its seeds more slowly. Over the winter, white-wings continued feeding mostly on Red Spruce, but by spring their intake rates declined so that it was more profitable to feed on the many opening cones of Black Spruce. Black Spruce has thick and surprisingly rigid scales, so that even though all the cones open by spring, seed shedding occurs over a protracted period with some seeds remaining in cones

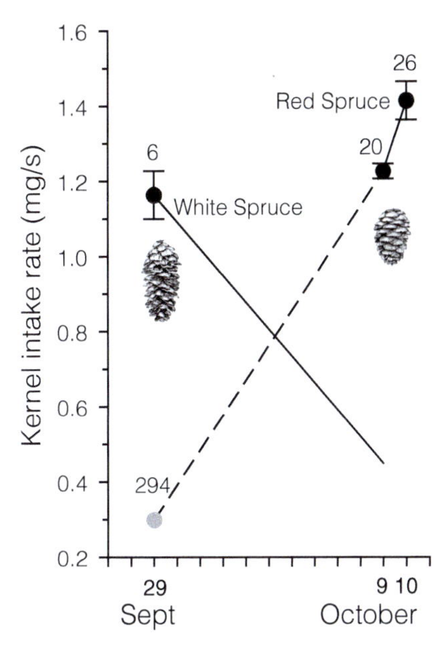

FIGURE 2.5 White-winged Crossbills *Loxia l. leucoptera* in Fundy National Park, New Brunswick, Canada in 1988, initially foraged exclusively on White Spruce *Picea glauca* in late September, but then by 9 October switched to foraging exclusively on opening Red Spruce *P. rubens* cones because they provided higher intake rates. Black circles represent data (means) from Fundy National Park, whereas the gray circle represents the mean from a site to the north without White Spruce and where crossbills foraged exclusively on Red Spruce. The projected decrease in intake rate on White Spruce (descending solid line) is based on similar decreases found in the fall for crossbills foraging on White Spruce elsewhere (Benkman 1987a). The numbers represent the number of foraging bouts (cones) for which I recorded intake rates, and the error bars around the means represent standard errors. From Benkman (2024).

for years. I refer the reader to Benkman (1987a) for data showing switches between these and other conifers as relative intake rates switched similarly to those in Figure 2.5. However, the gaps in the data (because I was not in the field) between when crossbills foraged almost exclusively on one conifer and then switched to another as relative intake rates shifted were generally of a month or two. A similar gap in my data occurred when white-wings in New Brunswick switched from foraging on Red Spruce in mid-February to Black Spruce in mid-April in 1989.

Such switches from feeding initially on abundant seeds in thinner-scaled cones to seeds in thicker-scaled cones that either open later or hold their seeds longer are common wherever there are multiple species of conifers. For example, in the Rocky Mountain region, larger-billed crossbills switch from feeding on seeds in the relatively thin-scaled Engelmann Spruce *Picea engelmannii* cones to feeding on thicker-scaled pine as spruce seeds decline in profitability. However, during particularly large spruce seed crops, seeds are sometimes held until the following summer and even large-billed crossbills sometimes rely on spruce for most if not all of an entire year. In northern Europe, Red Crossbills feed on Norway Spruce *Picea abies* starting in late summer and then switch to the thicker-scaled Scots Pine *Pinus sylvestris* in late spring the following year when its cones open (Juutinen 1953; Haapanen 1966; Newton 1972). The extremely deep-billed Parrot Crossbill *Loxia pytyopsittacus*, however, is more likely to rely on seeds in Scots Pine year-round (Newton 1972; Marquiss and Rae 2002; Summers et al. 2010). In Scotland, similar switches occur in Red Crossbills; however, they also feed on non-native conifers such as larch, Sitka Spruce *Picea sitkensis* and even Lodgepole Pine *Pinus contorta* in the fall and winter, and then often switch to Scots Pine as its cones open in the spring (Marquiss and Rae 1994, 2002; Summers et al. 2010).

Timing of breeding

> In crossbills ... food probably is both the proximate and ultimate factor producing breeding.
>
> (H.B. Tordoff and W.R. Dawson 1965: 420)

Finding and tracking fluctuating conifer seed crops is often a challenge, but it also provides opportunities for crossbills to breed repeatedly during the year, including during extreme conditions. Indeed, it is quite a sight to witness crossbills nesting in winter when days are short and often rather cold. I have indelible memories of stunning red, black and white male White-winged Crossbills singing while flying over snow-ladened spruce when temperatures hovered at –20°F (–29°C). Equally remarkable was seeing fledgling white-wings being fed regurgitated boluses of spruce kernels by their parents when I was struggling to stay warm as morning temperatures approached –40°F (–40°C). In the example in the previous section of white-wings feeding in sequence on three species of spruce, nesting occurred three times including when morning temperatures approached –40°F. They nested in late summer while feeding on White Spruce, then again in January and February while feeding on Red Spruce, and finally in April while feeding on Black Spruce. These three periods—late summer, late winter and spring—are when white-wings generally nest, at least if they find sufficient seed (Benkman 2020; Fig. 2.6). Most other crossbills tend to breed during one or more of these time periods (Benkman and Young 2020).

In North America, crossbills most commonly nest in July and August. This coincides with the ripening of seeds, such that fledged young are able to feed on the readily accessible seeds in opening cones. I suspect that at least one form of crossbill nests in late summer on nearly all conifers regularly fed upon by crossbills in North America. Exceptions include Red Pine *Pinus resinosa* in the northeast, as seeds in its hard developing cones seem to be utilized rarely if at all during summer, and Black Spruce in the northern boreal forests due to the difficulty of extracting seeds from its closed cones (see Fig. 3.1). Another exception is

FIGURE 2.6 A female Scottish Crossbill *Loxia scotia*) regurgitating seed kernels to nestlings. Crossbills rely nearly exclusively on conifer seeds, including while nesting. Crossbills generally lay three eggs and, as in the photo, have three nestlings. Droppings from the nestlings increasingly litter the nest rim as nestlings grow. Photograph taken by David Whitaker in Shin Forest, East Sutherland, Highlands of Scotland on 28 March 1982.

the Cassia Crossbill *Loxia sinesciuris*, which relies on seeds in a highly serotinous population of Lodgepole Pine whose cones do not open on a seasonal basis. Nesting is less regular in late summer in Europe than in North America, presumably because most conifers in Europe besides larch open in winter or spring. I ignore true fir (*Abies*) because crossbills forage on fir seeds only infrequently. Fir seeds are readily accessible as their cones disintegrate in the fall, but the seeds have resins that likely interfere with protein absorption by crossbills (Benkman 1987a).

Nesting often occurs in late winter and spring in North America, following large seed crops and when seeds remain in sufficient numbers in the cones. Although some populations of crossbills in Europe regularly nest in winter, such as those in the high-elevation Mountain Pine *Pinus uncinata* woodlands in the Pyrenees (Clouet 2000), the most regular nesting period in Europe is spring. This corresponds to when the cones of most native conifers open (Fig. 2.2); cones of Mountain Pine begin opening earlier in winter.

As might be expected given the fairly regular breeding of crossbills in mid to late summer, at least some crossbills exhibit seasonal cycles of gonadal development then. Tom Hahn studied the breeding physiology of Type 3 Red Crossbills in the Pacific Northwest—the call type Anna Lindholm and I studied in the bill trimming experiments described in Chapter 1—and discovered that gonads begin to grow and develop (termed recrudescence by physiologists) in early summer even before seed crops are located (Hahn 1998). This makes adaptive sense, because this is when the readily accessible seeds in Western Hemlock *Tsuga heterophylla* cones are maturing, and gonadal recrudescence ensures breeding readiness upon locating large hemlock seed crops.

Gonad recrudescence prior to locating seed crops is less evident at other times of the year. One reason why crossbills might not exhibit gonad recrudescence is that conditions suitable for nesting are not regular and outside of summer crossbills often settle weeks if not months prior to nesting. For example, nesting in winter and early spring generally requires both large cone crops and weather conditions conducive for the retention of many seeds in cones. As a result, seed crops that are adequate for nesting in summer are often inadequate for nesting in winter and early spring. In addition, the long interval between arrival and nesting in winter and spring allows time for crossbills to be stimulated into breeding condition when seed is plentiful. This is unlike in summer when crossbills may begin nesting within days of arriving, such that prior gonad recrudescence is particularly advantageous.

While measuring intake rates during my first year of fieldwork (1982–3), I noted that breeding commenced when intake rates were high or increasing and ceased after intake rates declined. This spurred me to examine whether breeding occurred consistently when energetic intake exceeded the energetic demands of reproduction. I perused studies on the energetics of reproduction in birds and used information and allometric equations from the literature to estimate the daily energy demands for a female crossbill to form eggs. Egg production for small songbirds like crossbills potentially limits the timing of nesting, but it is not the most energy-demanding stage. The most energy-demanding stage for crossbills instead falls three weeks later at the beginning of the nestling period. During the first five days, female crossbills usually remain at the nest brooding the nestlings—usually three of them (Fig. 2.6)—which are initially naked and unable to effectively thermoregulate. At this time, the male alone is responsible for feeding the nestlings, the female and himself (Bailey et al. 1953; Newton 1972; Nethersole-Thompson 1975). I also estimated the intake rates needed to produce eggs if protein alone was limiting, and concluded that egg production was much more likely to be limited by energy than protein requirements.

I then estimated the necessary intake rates for a female to form eggs (egg formation threshold) and a male to feed its family (nestling threshold) given the number of daylight hours, average nighttime and daytime temperatures, and assumptions of how much time could be spent feeding given other mutually exclusive activities, like flight (Benkman 1990). These estimates are shown in Figure 2.7 for white-wings in southern Canada, where I made most of my observations. These two sets of estimates or thresholds do not take into account seasonal changes in intake rates. They simply represent the projected intake rates necessary to meet the demands of egg formation and raising nestlings. However, intake rates regularly increase in late summer, decline in fall and winter, and for white-wings increase in spring. Because of these seasonal changes in intake rates, I also estimated what intake rates would be necessary when producing eggs that would later, based on the intake rate trajectory, be sufficient for the nestling stage. That is, do both current intake rates and the trajectory of intake rates influence nesting decisions? This so-called future nestling threshold is higher than the nestling threshold when intake rates are declining, and lower than the nestling threshold when intake rates are increasing (Fig. 2.7).

Although breeding obviously requires egg production, nearly all populations had intake rates sufficient for this, implying that the ability to form eggs alone

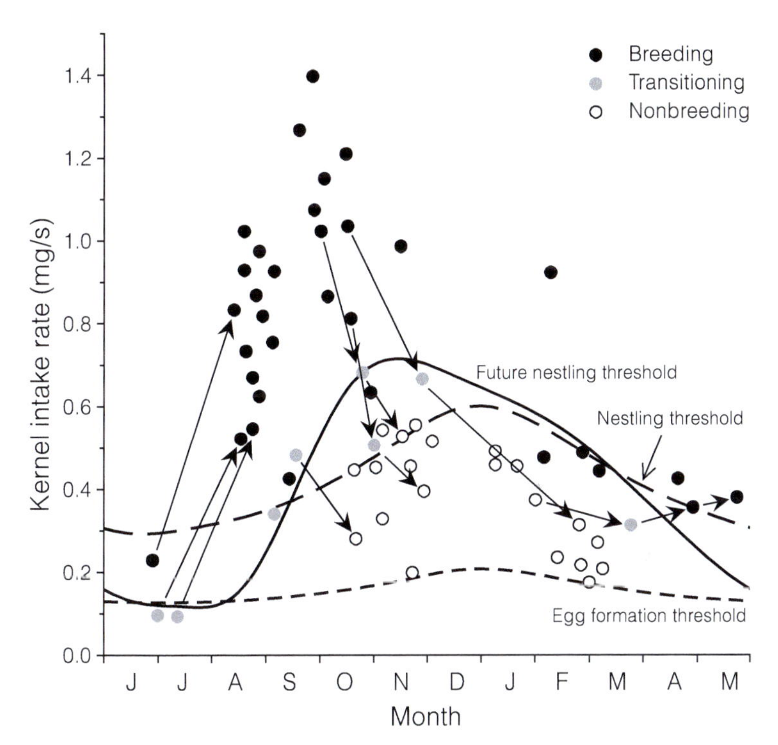

FIGURE 2.7 Kernel intake rates of White-winged Crossbills *Loxia l. leucoptera* in relation to breeding status and the three potential thresholds for breeding. Arrows with black tips connect successive intake rates for a given population; not all the successive intake rates from the same population are connected by arrows. Intake rates increase in summer, decrease in autumn, and can increase in spring when switching to forage on opening Black Spruce cones. These arrows also show that the future nestling threshold, which considers the trends in intake rates over time, is often crossed when populations transition from nonbreeding to breeding and from breeding to nonbreeding. Transitioning in early July and late March represents populations just prior to nest building (paired and singing), whereas transitioning in September through November represents only one or a few pairs with fledged, dependent young with no evidence of earlier stages of nesting. Intake rates were based on 4,559 foraging bouts and 44,315 seeds consumed. Modified from Benkman (1990).

rarely limits the timing of breeding. Similarly, white-wings did not always wait to begin nesting or continue nesting when their intake rates exceeded the nestling threshold (Fig. 2.7). Instead, white-wings seemingly nested whenever their intake rates, based on intake rate trajectories, would be sufficient for the even more demanding early nestling stage (Fig. 2.7; future nestling threshold) when the male generally provides all the food for his family (Benkman 1990). I had much less data for Red Crossbills, but the pattern was similar. Crossbills often began nesting in summer when intake rates were relatively low and just sufficient for egg formation, because the predictable seasonal increase in intake rates ensured they would be prepared for the most energy-demanding early nestling stage three weeks later. Conversely, several months later in the fall crossbills stopped nesting, even though immediate intake rates were sufficient for the nestling stage and were much higher than when nesting was initiated in summer (Fig. 2.7). Nesting ceased apparently because the declines in intake rates were such that intake rates would be inadequate three weeks later for the early nestling stage.

One limitation of these data was that I simply recorded the presence of breeding behavior and nesting activities, not the frequency of occurrence among individuals. If for example only one pair was found breeding (e.g., constructing a nest or feeding nestlings), I would record the population as breeding. My impression is that only a small proportion of individuals began nesting in autumn after September, which could be due both to the need to molt and the fact that fledging young at the beginning of winter is likely an inopportune time. Regardless, the results (Fig. 2.7) indicate that seed profitability drives the timing of breeding, and that crossbills behave as if they are able to forecast when future intake rates will be sufficient to raise nestlings. However, the fact that not all individuals were in synchrony indicates that although intake rates limit their timing of breeding, other considerations like the recency and success of previous nesting attempts, the need to molt and the prospects of the young surviving likely act to modulate their responses to breeding opportunities.

In sum, the seed ripening and shedding phenology of conifers affects conifer use, movements and timing of breeding by crossbills. Crossbills track the seasonal variation in seed profitability among conifers both within and among habitats, and opportunistically nest nearly whenever they are likely able to raise and fledge young. The clear relationships between crossbill behavior and ecology and the profitability of seeds underscores Hutchinson's (1959) emphasis on 'food relations'. However, to better comprehend some of Hutchinson's 'higher intricacies', we need to drill down further to understand how variation in bill structure among crossbills and cone structure within and among conifers influences crossbill conifer use.

Chapter 3

How Cone and Bill Structure Shape Conifer and Habitat Use

> In the continual play of factors of existence consequent upon the continual changing of general and local environments, birds, by reason again of their endowment of motility, of sensitiveness, of extreme alertness, are able quickly to make adjustments.
>
> (J. Grinnell 1931: 30–1)

Not only do intake rates drive crossbill preferences for different conifer species and motivate long-term movements and breeding, but these rates also drive the differential use by crossbills of trees within a conifer species, specialization within populations of crossbills and, ultimately, crossbill diversification. But first, I want to consider cues other than food that crossbills might use when choosing habitats. One cue is simply the occurrence of other individuals of their own species (so-called conspecifics). Attraction to, rather than repulsion by, conspecifics is aptly termed conspecific attraction (Stamps 1988). If conspecifics thrive in a habitat, then it is likely that a similar nearby habitat will allow them to thrive given that competition is limited. Likewise, individuals might seek to settle in a location where conspecifics bred successfully (Doligez et al. 2002). Cues of nesting success such as evidence of recent fledglings and their condition might be used to choose future breeding sites.

I believe crossbills, especially solitary crossbills, are commonly attracted to conspecifics especially when foraging. Indeed, the benefits of conspecific flocking are the focus of Chapter 6. Yet I don't believe information reflecting past breeding success is particularly useful to crossbills, given their flexible breeding timing and how much they move. As discussed in Chapter 2, 'Crossbill responses', most crossbills are unlikely to occupy let alone breed in the same location in successive years. Moreover, dynamic changes in cone-seed resources mean that breeding success in the previous months might not be indicative of subsequent breeding success at the same location. The need for continual sampling and evaluation of the environment through foraging was evident in the summer of 1988. This is when I began a project on White-winged Crossbills *Loxia l. leucoptera* in the Laurentian Mountains north of Quebec City, as white-wings sang and settled to breed in Tamarack *Larix laricina* forests where they had bred successfully in previous summers (in 1982 and 1984 during my dissertation fieldwork). The cone crop seemed adequate, but within several weeks the singing stopped and the crossbills departed, apparently in response to cone insects (*Hylemya laricicola* larvae) devouring much of the developing seed crop. The most reliable way to detect such a deterioration in the seed crop is by foraging on it, so this local decline in future food was undoubtedly obvious to those departing birds.

But are intake rates the only measure of habitat quality to a crossbill? Crossbills for example prefer more open woodlands over dense, closed forests. This preference was documented by Ron Summers and Robert Proctor (Summers and Proctor 1999) in Scots Pine *Pinus sylvestris* woodlands in Scotland. However, seed accessibility might also explain the preference by crossbills for more open woodlands, because Scots Pine in open woodlands had smaller cones with thinner scales and more accessible seeds than those on pine growing in denser stands (Summers and Proctor 1999; Summers et al. 2010). An additional reason for preferring open forests, related to an intake rate advantage, is the greater abundance of cones on open grown trees and perhaps less competition for cones from tree squirrels that generally occupy more closed forests (Mezquida and Benkman 2014). Nevertheless, a preference for open over closed woodlands seems general to crossbills, such as in the Mountain Pine *Pinus uncinata* forests in the Pyrenees (Mezquida and Benkman 2014) and in Scots Pine forests in Spain where tree squirrels were scarce and there was no evidence that cone structure differed in a habitat-specific manner (Mezquida and Benkman 2021).

The preference by crossbills for open woodlands could act to reduce predation risk faced by the birds (Benkman 1987a; Summers and Proctor 1999; Mezquida and Benkman 2014, 2021). Crossbills forage on cones near the ends of branches, often hanging upside down where they would seemingly be vulnerable to an ambush predator like a Sharp-shinned Hawk *Accipiter striatus* in North America or a Eurasian Sparrowhawk *A. nisus* in Europe (Petty et al. 1995). Greater spacing between trees should enable crossbills to detect from farther away an approaching *Accipiter* that uses cover to conceal its approach (Newton and Marquiss 1982), allowing crossbills to evade predation (Kenward 1978) by for example 'freezing' before being detected or by flying away.

Despite watching crossbills in the field for thousands of hours, I have never observed a successful attack by a predator on a crossbill. Though a student of mine, Trevor Fetz, who conducted foundational field research on Cassia Crossbills *Loxia sinesciuris* in the South Hills, had this rare opportunity. The crossbill was on the ground drinking water from a puddle in a dirt road when a Sharp-shinned Hawk swooped in and captured it. Crossbills seem to be especially vulnerable when they go to the ground for water or grit and salts, the types of locations where we commonly use mist nets to capture them. In fact, a Sharp-shinned Hawk frequented our netting site in the South Hills, Idaho in 2017, where Cassia Crossbills were flying to the ground at the base of a cabin to consume salts that had leached from the cabin's mortar. When we arrived in late May, we found multiple piles of feathers where birds including crossbills had been plucked by the hawk. However, the apparent rarity at which adult crossbills are preyed upon when not near the ground makes it difficult to quantify whether predation influences habitat use while foraging in trees.

Cone structure

Habitat structure aside, a crossbill's decision to use a particular habitat is strongly influenced by the profitability of seeds and the likelihood that crossbills can be supported by cone-seed for an extended period, as discussed in Chapter 2, 'Crossbill responses'. A major factor influencing the profitability of seeds to crossbills is cone structure. For example, thinner cone scales provide less resistance to crossbills using their mandibles to bite between and spread apart the scales (see Fig. 1.7). The increase in time to access and extract a seed as scale thickness increases is illustrated in Figure 3.1, where time per seed increases from the especially thin-scaled White Spruce *Picea glauca* (0.8 mm scale thickness) to the Red Spruce *P. rubens* (1.2 mm) and finally Black Spruce *P. mariana* (1.5 mm), with the last having especially rigid scales (Benkman 1987b). More seeds per cone reduce both the time to access seeds (Figs 2.4 and 3.1) and how often crossbills need to move to another cone. More cones with more seeds means seeds will remain abundant and intake rates will stay higher for longer. Larger seeds with larger kernels provide more energy; although larger seeds are generally associated with larger cones with larger, thicker scales that impede seed access.

Even more subtle differences in cone structure than those between different species of conifers are important to crossbill habitat use. This became evident after Tom Parchman—a former graduate student who is now a professor at the University of Nevada at Reno—and I analyzed our data from a trip to Greece. In early November 2008, we were in Greece searching for large cone crops of European Black Pine *P. nigra*. We were seeking to measure natural selection exerted by crossbills on cone traits, to further understand geographic variation in cones in relation to variation in the occurrence of crossbills and tree squirrels. We located an area with a large cone crop in the Pindos Mountains. Arriving in the afternoon, we walked among the pines and found evidence of crossbill-opened cones that had been dropped beneath trees (Fig. 3.2B). We spent the next few days counting the number of cones on the trees (nearly all untouched by foraging crossbills), and those that had fallen to the ground after the crossbills had removed the cones from branches and foraged on them. The number of foraged-on cones divided by the total number of cones was our measure of the proportion of the tree's seed crop eaten by crossbills. We also collected closed cones from the canopy so that we could characterize the traits of the cones from each tree, to relate seed predation by crossbills to cone traits as a measure of natural selection exerted by crossbills. I focused on cone counting, while Tom, the stronger of us, used a nine-meter pole with a clipper to collect cones from the canopy (Fig. 3.2A).

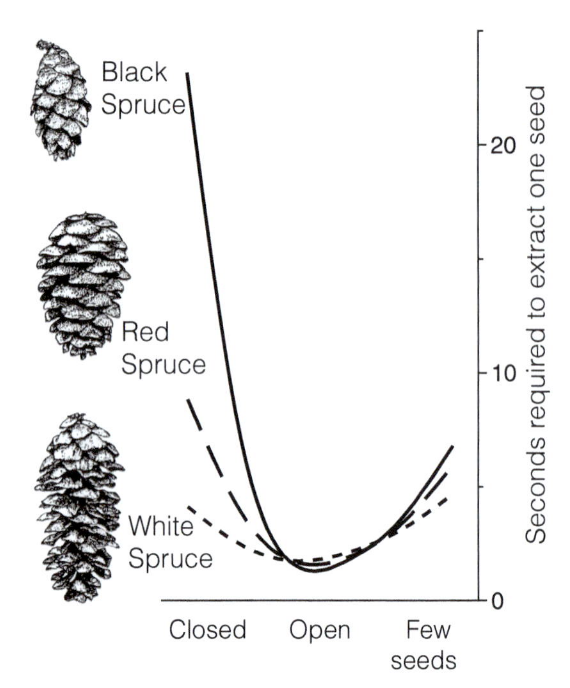

FIGURE 3.1 The time taken by captive White-winged Crossbills *Loxia l. leucoptera* to extract seeds from cones of three species of spruce in relation to cone phenology—from when cones are closed, to open cones full of seeds, to after they have shed most of their seeds. The increase in time to extract seeds when cones are closed or after most seeds are shed was greatest for Black Spruce *Picea mariana*, which has the thickest (1.5 mm), most rigid scales, and least for White Spruce *P. glauca*, which has the thinnest scales (0.8 mm). Time per seed was measured for six stages in the phenology and a spline was fit to the means. Modified from Benkman (1987b). Cone images from Benkman (1987b) and crossbill from Parchman and Benkman (2002). Size of crossbill is not scaled in proportion to the cones.

FIGURE 3.2 (A) Tom Parchman using an extension pole with a clipper at the end to collect closed cones from a European Black Pine *Pinus nigra* in the Pindos Mountains, Greece in November 2008. (B) Crossbill-opened cones that we piled at the base of a tree after we counted them.

After spending several days quantifying seed predation, we were a bit perplexed to find so few crossbills and so little seed predation in a seemingly good habitat filled with cones. Before we left, we decided to search elsewhere to see if crossbills were similarly uncommon throughout the general area. We drove north along the ridge to another large area of pines. We got out of the car in the evening fog. It was late enough that crossbills would have gone to roost, but not too late to see crossbill-discarded cones littering the ground under many of the trees. The next day we started surveying this new stand, which confirmed the abundance of foraging crossbills. Yet for the entire field period we remained puzzled about why two such similar pine stands three miles (5 km) apart (Fig. 3.3) differed so dramatically in the occurrence of crossbills. The difference was notably large, as the proportion of the seeds eaten was 12.4 times greater at the second site.

Not until the cones were measured back in Wyoming (I had moved from New Mexico State University to the University of Wyoming in 2004) did we figure out why crossbill abundance differed between the two sites. Both the average cone mass and average scale thickness were greater in the area where crossbills were scarce than in the area where they congregated and foraged (7% [1.08 gm] and 4% [0.13 mm] greater, respectively). Why cones on average differed between the two nearby locations is unknown; however, the foraging behavior of crossbills acts to reduce such variation between locations over time (Benkman and Parchman 2013). By disproportionately eating seeds from trees with smaller, thinner-scaled cones at the site with more of such cones, future generations at this site will be increasingly dominated by larger, thicker-scaled cones and converge with those at the other site until cone use and cone traits equilibrate between the two sites.

The consistent avoidance by crossbills of trees having larger, thicker-scaled cones within forests in the Pindos Mountains (Benkman and Parchman 2013) is repeated in other locations and pines (Fig. 3.4; Benkman and Mezquida 2015). Moreover, we found similar results in aviaries. Crossbills have their highest intake rates when foraging on small, thin-scaled cones of Lodgepole Pine *Pinus contorta* (Benkman 1999), and preferentially forage on such cones of both Ponderosa Pine *P. ponderosa* (Parchman and Benkman 2008) and Black Pine (Benkman and Parchman 2009). The consistent cone/tree preferences presumably reflect crossbills' unfailing quest for cones providing the highest intake rates (Chapter 2, 'Crossbill responses') and their stereotypic foraging behavior (Chapter 1, 'Getting seeds from cones'). Knowing these preferences is critical for understanding coevolution between crossbills and conifers, which is the focus of Chapter 5.

In addition to thicker cone scales, spines on the scale tips (Fig. 3.5) impede crossbills (Coffey et al. 1999). What sparked my interest in addressing the effect of spines on foraging crossbills were differences between the cones of two sister species of pines: Jack Pine *P. banksiana* and Rocky Mountain Lodgepole Pine *P. contorta* var. *latifolia* (Jin et al. 2021). Both of these pine species have extensive distributions in North America and produce similarly sized hard, woody serotinous cones that remain closed and hold their seeds for multiple years. However, Lodgepole Pine

FIGURE 3.3 Photographs of European Black Pine *Pinus nigra* at the two field sites in the Pindos Mountains, Greece. Crossbills were relatively scarce at (A), whereas they were common three miles (5 km) away at (B). Both sites had many open stands of pine that crossbills generally prefer.

FIGURE 3.4 Eduardo Mezquida found that crossbills preferentially forage on Mountain Pine *Pinus uncinata* trees in the Pyrenees that have thin-scaled cones (Mezquida and Benkman 2010, 2014). The cone on the left is from a tree with many cones shredded by foraging crossbills, while the cone on the right with much larger scales is from an adjacent tree with no cones foraged on by crossbills. Photograph taken in the Vercors Massif, France on 6 September 2009.

FIGURE 3.5 Pine cones with spines. (A) Close-up of an open cone of Mountain Pine *Pinus uncinata* in the Pyrenees, (B) an open cone of Table Mountain Pine *P. pungens*, (C) close-up of the spines on a closed cone of Rocky Mountain Lodgepole Pine *P. contorta* var. *latifolia*, (D) a closed cone of Rocky Mountain Ponderosa Pine *P. ponderosa* var. *scopulorum* and (E) close-up of an open cone of Rocky Mountain Ponderosa Pine. The Table Mountain Pine cone (B) is secured to a dowel for foraging experiments to test whether spines slow the rate at which crossbills extract seeds from cones.

cones have spines (Fig. 3.5C), whereas Jack Pine cones are smooth. If spines were an important defense against seed predators like tree squirrels that feed nearly exclusively on seeds in closed cones, then we would expect Jack Pine to have spines on its cones too. Why then does Lodgepole Pine but not Jack Pine have spines on its cones? We think it's related to the duration of time seeds are held in open cones: seeds are shed quickly in Jack Pine, but more slowly in Rocky Mountain Lodgepole Pine (Critchfield 1985). The longer seeds are held in open cones, the greater the need to deter animals like crossbills that commonly forage for such seeds. Many pines foraged on by crossbills have spines at the tips of their scales that are directed into the gap between open cone scales, where crossbills reach to extract seeds (Fig. 3.5). These spines also make it difficult to perch on the cones.

I followed up these observations by tabulating data from the literature on the occurrence of spines, and the duration of time that different pine species held seeds in closed and open cones. Excluding white or soft pines (subgenus *Strobus*), because none of them have spines, there was a strong tendency for species that held their seeds for an extended duration in open cones to have spines. In contrast, there was no tendency toward spines in species that held seeds for extended durations in closed cones. These general patterns are consistent with the pattern found in Jack and Lodgepole Pines. What we needed to do was conduct experiments with crossbills to test whether spines were an impediment to their foraging. Around this time, Kimberly Coffey joined my lab. Kim was a graduate student with an interest in both plants and birds. Moreover, she had lived in the southern Appalachian Mountains, where she was familiar with and intrigued by the impressive spines on the cones of Table Mountain Pine *P. pungens* (Fig. 3.5B).

Kim's thesis research involved measuring the intake rates of captive Red Crossbills *Loxia curvirostra* foraging on closed and open cones of Rocky Mountain Ponderosa Pine *P. ponderosa* var *scopulorum* (Fig. 3.5D,E), including cones from which she had removed all the spines. She also repeated the experiments using Table Mountain Pine cones (Fig. 3.5B) she had gathered in North Carolina. In these trials, Kim used only open Table Mountain Pine cones, because the crossbills could not remove seeds from the mature closed cones. Kim found that on closed Ponderosa Pine cones, the presence of spines did not impede crossbills consistently. Two of five crossbills took longer to remove seeds from cones with spines, whereas three of five took longer to remove seeds from cones without spines. In contrast, all seven crossbills (two of the seven wouldn't even forage on the closed cones) took longer to remove seeds from open cones with spines than from those without spines. The occurrence of spines on open Ponderosa and Table Mountain Pine cones increased the amount of time crossbills required to remove each seed by 18% and 34%, respectively (Coffey et al. 1999). The difference between the increases in foraging time on the two pines is presumably related to the larger spines on the cones of Table Mountain than on those of Ponderosa Pine (Fig. 3.5B,E).

Kim's experimental results are consistent with the hypothesis that spines are important for impeding crossbills, and presumably numerous less specialized birds and small mammals, in extracting seeds from between open scales. Moreover, the orientation of the spines on open Ponderosa Pine cones (Fig. 3.5E) suggests a deterrent function. Indeed, the spines are often close to the eyes of foraging crossbills as they are trying to extract seeds. In fact, I suspect that spines generally impede crossbills foraging on closed cones too. Crossbills foraging on closed Ponderosa Pine cones (Fig. 3.5D) sometimes spend time tearing off spines. Nevertheless, Kim's inability to detect an effect of spines on closed cones suggests that unmeasured cone traits dominate and hence mask the effect of spines. This is consistent with the data in Figure 3.1: differences in scale thickness have a massive effect on the time per seed for crossbills foraging on the closed cones of different species of spruce, but only a minor effect once the cones open. A dominant spine deterrent effect for open cones only is further consistent with the association of spines with the duration of time that seeds are held in open but not closed cones, mentioned earlier.

We were fortunate that Brook Milligan—an evolutionary biologist on the faculty in our department at New Mexico State University—was on Kim's thesis committee. Brook had recently developed phylogenetic methods that were appropriate for inferring the origin and evolution of traits using data from multiple species. He offered to analyze the spine and seed retention data using a pine phylogeny from the literature that included the 21 pine species for which Kim and I had found such data. Brook's analyses strongly supported our inference that the evolution of spines is related to seeds being held for longer durations in open but not closed cones (Coffey et al. 1999). It is always reassuring when results from analyses of independent datasets—in this case comparative and experimental data—are congruent.

The interaction between bill and cone structure

The accessibility of seeds, especially those in closed cones, depends also on the match between the bill of the crossbill and the structure of the cones. An effect of differences in bill and cone structure is obvious from comparisons of the use of conifers by different species of crossbills in the same habitat. During my dissertation I measured intake rates and conifer use of the small, slender-billed White-winged Crossbill and the larger-billed Red Crossbill (Fig. 3.6). Both species foraged on the readily accessible seeds in the small, thin-scaled closed White Spruce cones during summer (Benkman 1987a), but as seeds and cones matured, Red Crossbills switched to forage on the larger-seeded and thicker-scaled Eastern White Pine *Pinus strobus* in early fall (see open cones in Fig. 3.6). This switch by Red Crossbills occurred as the White Pine cones opened and their seeds became increasingly accessible, so that intake rates on pine exceeded that on spruce. White-wings, however, continued to forage for seeds in the small, thin-scaled cones of White

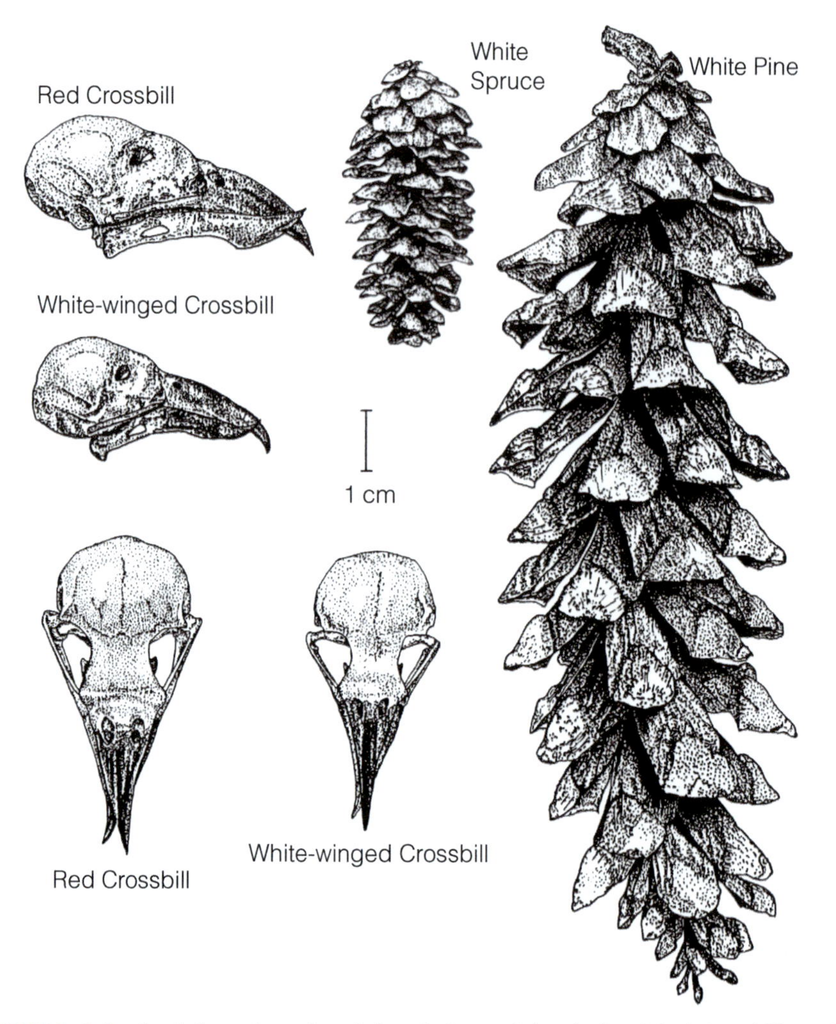

FIGURE 3.6 On the left are lateral and dorsal views of the skulls of Red Crossbills *Loxia curvirostra* and White-winged Crossbills *Loxia l. leucoptera*, and on the right are cones of White Spruce *Picea glauca* and Eastern White Pine *Pinus strobus*, all drawn to scale. Drawings from Benkman (1987b).

Spruce. The seeds in the large-scaled cones of White Pine were more difficult for white-wings to access, and they struggled with their small slender bills to husk the large seeds and not drop seed kernel.

Performance-based habitat choice

Individuals within the same crossbill population can differ among themselves in habitat or conifer use, too. This would be expected if larger-billed individuals have higher intake rates on one cone type, but on another cone type the reverse occurs and smaller-billed birds are more efficient. Then individuals might sort themselves by habitat based on cone type, and thereby potentially use different forests. Such sorting by bill size to match the habitat providing the highest reward or fitness has been termed 'matching habitat choice' (reviewed in Edelaar et al. 2008) or 'performance-based habitat choice' (Munar-Delgado et al. 2024). It is expected to occur in animals like crossbills that have strong powers of flight allowing them to sample multiple habitats, and when habitats are heterogenous (Bolnick and Otto 2013). Spatial variation in the occurrence of different species of conifers and even spatial variation within a conifer species (as in the Pindos Mountains) are good examples of how habitats might vary spatially. However, determining whether individuals within a population make such adaptive choices is an empirical challenge. Examples from the wild therefore are few. A particularly elegant experiment by Munar-Delgado and colleagues (Munar-Delgado et al. 2024) using Eurasian Tree Sparrows *Passer montanus* differentially and randomly assigned resource access to individuals. Most individuals moved to habitats where they had greater access to resources, and those that did move accordingly (and adaptively) had higher reproductive success.

Several examples from crossbills are consistent with performance-based habitat choice (Summers et al. 1996; Marquiss and Rae 2002; Siepielski and Benkman 2005; Gómez-Blanco et al. 2019). Marquiss and Rae (2002) for example found that only the largest-billed individuals (among a heterogeneous mix of individuals that likely included Parrot Crossbills *Loxia pytyopsittacus*, Scottish Crossbills *L. scotia* and Red Crossbills) remained for more than six months in woods dominated by the hard-coned Scots Pine on the Deeside in Scotland. Because smaller-billed individuals are less able to forage on closed Scots Pine cones (Summers et al. 2010), these smaller individuals likely moved on and settled in areas containing other conifers with softer cones (Marquiss and Rae 2002).

I tested for performance-based habitat choice using several sources of data, including those gathered by capturing and banding crossbills over multiple years in the South Hills in southern Idaho. Most crossbills captured were the resident Cassia Crossbill. However, we also caught individuals of two call types of Red Crossbill that commonly move through the South Hills in late spring and early summer during their yearly search for developing seed crops. Even though Cassia Crossbills have sufficient intake rates to raise young at this time, most Red Crossbills do not appear to stay for long. Instead, they seemingly pass through because of their inability to forage efficiently for seeds in Lodgepole Pine cones in the South Hills. These cones have especially thick scales as a defense against crossbills because of a coevolutionary arms race between the Cassia Crossbill and Lodgepole Pine, a subject I discuss in Chapter 5. Cassia Crossbills in turn have the largest bills of all crossbills regularly found north of Mexico, outside of Newfoundland. The Cassia Crossbill's large bill is an adaptation for foraging on these crossbill-defended cones, as indicated by their average bill depth (arrow in Fig. 3.7), approximating the optimal bill depth based on aviary foraging experiments (dashed curve in Fig. 3.7).

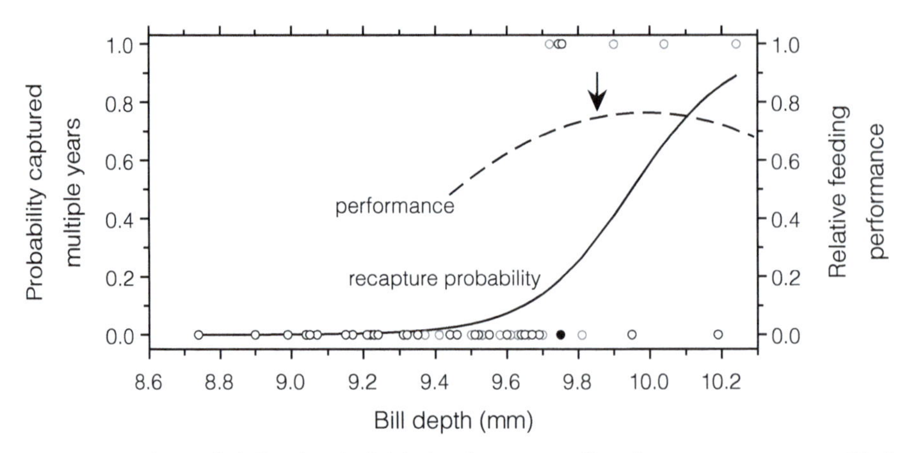

FIGURE 3.7 The probability that individuals of two nomadic call types or ecotypes of Red Crossbill *Loxia curvirostra* were captured in two or more years (solid line) in the South Hills, and relative feeding performance (dashed line) in relation to bill depth. The arrow represents the average bill depth of the resident Cassia Crossbill *Loxia sinesciuris*. The circles represent individuals that were either captured in only one summer (along the bottom) or during multiple summers (along the top). The dark circle along the bottom at 9.75 mm represents an individual photographed two years later 406 miles (650 km) to the east of the South Hills, consistent with the idea that most individuals that were captured during only one summer moved on rather than died. Modified from Benkman (2017a).

The two Red Crossbill call types that move into and through the South Hills—I refer to them as the Lodgepole Pine and Ponderosa Pine ecotypes because they are adapted to these two conifers, as we will see in Chapter 4—average smaller than that of the resident Cassia Crossbill: 9.33 mm and 9.56 mm in bill depth respectively, versus 9.85 mm for the Cassia Crossbill. These smaller-billed ecotypes have difficulty accessing seeds in the Lodgepole Pine cones in the South Hills, as indicated by the relationship between feeding performance (ability to meet their estimated energy demands) and bill depth (dashed curve, Fig. 3.7). The relatively small-billed Lodgepole Pine ecotype is adapted to Lodgepole Pine outside the South Hills (Benkman 1993a), where the pine is not engaged in a coevolutionary arms race with crossbills and the cones have thinner scales (Benkman 1999).

Even though both the Lodgepole Pine and Ponderosa Pine ecotypes average smaller than the Cassia Crossbill, there is considerable overlap in bill depth among them, with some individuals of the two ecotypes equaling or exceeding the average bill depth of the Cassia Crossbill. The bill depths of the individuals of these ecotypes that we initially captured in 2009, 2010 or 2011 are shown as circles in Figure 3.7. Assuming performance-based habitat choice, those individuals that remained in the South Hills for multiple years (circles along the top axis) should have bill depths approximating the average bill depth of the Cassia Crossbill. Whereas those that were captured for one summer only (circles along the bottom axis), and presumably moved on, should have had smaller bills because of their poor feeding performance. That is indeed what we found (Fig. 3.7): none of the individuals with bill depths less than 9.7 mm were recaptured in multiple years, while our probability of recapturing individuals in multiple years increased with further increases in bill depth (solid curve in Fig. 3.7).

An alternative explanation for the pattern in Figure 3.7 is differential mortality (i.e., natural selection) rather than habitat choice. Individuals that were not detected in a second year could have died rather than moved on. Discriminating between these two possibilities is one of the challenges of documenting performance-based habitat choice (Edelaar et al. 2008). One individual of the Ponderosa Pine ecotype that we captured in 2009 was later photographed 406 miles (650 km) to the east near Casper, Wyoming in 2011. This individual is indicated by the dark circle in Figure 3.7. It is one of only two adult Red Crossbills that we have captured and color-banded in the South Hills over 21 summers that have been found elsewhere (out of the 258 adult individuals of the Lodgepole Pine and Ponderosa Pine ecotypes that we color-banded). The second individual was also a Ponderosa Pine ecotype. He was initially captured in June 2018 and photographed two and a half months later 144 miles (230 km) to the northwest. These two birds indicate that the brief seasonal occurrence of most Lodgepole Pine and Ponderosa Pine ecotypes in the South Hills is because they move on rather than die there.

The individuals of the Lodgepole Pine and Ponderosa Pine ecotypes that became resident further illustrate crossbill opportunism and behavioral flexibility in response to variation in the profitability of seeds. Individuals of the Lodgepole Pine and Ponderosa Pine ecotypes and other ecotypes wander until they find plentiful, accessible seeds and then remain until they are unlikely to be supported much longer. Generally, crossbills leave within several months to a year, but in the South Hills seeds remain plentiful year after year and well-matched crossbills continue to stay. This illustrates how easy it is for crossbills, which are behaviorally flexible in their propensity to move and settle (Newton 1972, 2006; Benkman 1987a, 2017a; Koenig and Knops 2001), to switch from being nomadic, wandering widely one to several times a year in search of cone-seed, to becoming sedentary if conditions allow. Indeed, as Svärdson (1957: 327) noted when discussing nomadic movements of crossbills and other northern birds, 'a rich supply of food depresses their mobility and releases a sedentary habit'. These findings also align with Joseph Grinnell's assertion that 'the easiest thing they [birds] can do toward maintaining successful existence is to transfer their behavior from that of a population shifting annually to the condition of continual residence' (1931: 31).

Performance-based habitat choice, which I suspect is a crossbill's way of life, accelerates the rate at which crossbill populations diverge and adapt to alternative conifers (Armsworth and Roughgarden 2008; Edelaar et al. 2008; Bolnick and Otto 2013). Natural selection causes adaptation when individuals with bill structures well matched to their local environment are more likely to survive and reproduce than those whose bills are less well matched, resulting in an increase across generations in the proportion of individuals with well-matched bill structures in the population. Performance-based habitat choice complements and accelerates the action of natural selection, but without necessarily reducing the fitness of locally maladapted individuals if those individuals move and are able to locate resources on which they are well matched. Such habitat choice could act rapidly, resulting in different size classes of crossbills associating with different conifers on which they are best adapted. This divergence to specialize on different conifers—presumably as the result of both natural selection and performance-based habitat choice—is the subject of the next chapter.

Chapter 4

How and When Key Resources Favor Specialization

> The correlation between beak size and type of cone eaten could profitably be studied further in this genus (*Loxia*).
>
> (D. Lack 1944a: 269)

When I began research on crossbills, it was widely believed that each species of crossbill in northern Europe was adapted to extract seeds from a different species of conifer. The smallest-billed Two-barred Crossbill *Loxia l. bifasciata* is adapted to removing seeds from the small, thin-scaled cones of larch (especially two larch species, *Larix sibirica* and *L. gmelinii*, with similar cones that together occur across most of northern Eurasia); the medium-sized Red (Common) Crossbill *L. c. curvirostra* is adapted to the thicker-scaled cones of Norway Spruce *Picea abies*; and the massive-billed Parrot Crossbill *Loxia pytyopsittacus* is adapted to the hard and even thicker-scaled cones of Scots Pine *Pinus sylvestris* (Lack 1944a,b; Newton 1972). In northeastern North America, where I was planning to start my dissertation, and across North America more generally, less was known about the foraging associations of crossbills, and it was unclear whether each taxon of crossbill was adapted to a particular conifer or set of conifers. By the time I finished my dissertation, I was still uncertain whether one could conclude that Red Crossbills in the east were adapted to particular conifers, in part because I had limited observations of the different subspecies of Red Crossbills (Griscom 1937). Most of the Red Crossbill taxa were common only in the west. In addition, south of the boreal forest where Red Crossbills were most often found, the eastern forests had undergone substantial deforestation and subsequent reforestation, sometimes with non-native conifers. How this has and will play out remains to be determined.

In contrast, I had many opportunities to observe White-winged Crossbills *Loxia l. leucoptera* throughout most of the year. Moreover, white-wings generally occur in the boreal forest north of where Red Crossbills usually reside, and although substantial logging had and was occurring in the northern boreal forest, vast expanses of native conifers remained, and non-native conifers were not used in reforestation. Over the three years that I made extensive observations of white-wings, they fed mostly on three native conifers: White Spruce *Picea glauca*, Black

Spruce *Picea mariana* and Tamarack *Larix laricina*. I also observed white-wings feeding on Red Spruce *Picea rubens*, but this species occurs in only a small fraction of the range of white-wings, and in most years most white-wings probably do not encounter Red Spruce.

Based on my observations described in Chapter 2, white-wings search for large developing seed crops of White Spruce and Tamarack each summer, most commonly breeding while feeding on their seeds, and the vast majority of seeds eaten are from these two conifers. Given that so much of white-wing behavior and ecology is tied to White Spruce and Tamarack seeds, it seemed logical that white-wings would be adapted for foraging on seeds of one or both of these two conifers. Indeed, anyone who offered an opinion suggested that white-wings are adapted to White Spruce, the conifer on which they feed most (see Fig. 6.1).

When I started my fieldwork, I envisioned that white-wings might be adapted specifically for foraging on Tamarack. Tamarack is a larch (also called American Larch), and the similar Two-barred Crossbill is adapted to larch in Eurasia. Two observations, however, indicate that it would be a poor strategy to specialize on either Tamarack or White Spruce. First, the thin scales of these two conifers are prone to opening, causing the rapid release of seeds, especially from the pendulous White Spruce cones. In the late winter of 1983, a warm, dry spell was apparently enough to cause White Spruce cones that had reclosed over winter to open and shed most of their remaining seeds (the average number per cone declined from nearly 10 to only half a seed). This caused white-wings to emigrate even though a month earlier seed had been adequate to begin nesting (Benkman 1990). Only decades later did I learn of a key difference between Tamarack and one of the larches—*Larix gmelinii*—relied upon by the Two-barred Crossbill. *L. gmelinii* holds seeds through winter and does not begin shedding seeds until spring or summer, and some seeds are held in cones for three to four years (Abaimov et al. 1998). In contrast, and as I found out early on, Tamarack seeds are shed much more rapidly, such that only a few remain in the cones through the first winter. *L. gmelinii* and other common Siberian larches fed on by Two-barred Crossbills also dominate a vastly greater area than Tamarack (Schulte et al. 2022), which makes it more likely that good larch seed crops can be found every year.

Second, smaller finches such as Redpolls *Acanthus flammea* and Pine Siskins *Spinus pinus* are equally adept as white-wings at removing and husking seeds from dry, open or even partially reclosed White Spruce and Tamarack cones (Benkman 1987a). Because Redpolls and Pine Siskins have only half the body mass of a white-wing and hence have much lower daily energy requirements, these finches can meet their necessary intake more quickly. This allows them to survive on lower abundances of White Spruce and Tamarack seeds and thus outcompete white-wings by lowering (via consumption) the abundance of seeds below the number on which white-wings can survive. Such a scenario appeared to unfold while I was studying white-wings in the winter of 1984–5 in the Laurentian Mountains, Quebec. White-wings fed mostly on seeds in Tamarack cones from summer through fall in 1984, as in 1982. Unlike in 1982, large numbers of Redpolls moved in by November 1984 and foraged heavily on Tamarack cone-seed. As a result, seed abundance, and white-wing intake rates and abundance, declined sharply between October and November (Benkman 1987a). When I returned in February 1985, the few remaining white-wings foraged exclusively on Black Spruce seeds, while Redpolls continued to forage on Tamarack. In 1982–3, when Redpolls were absent, white-wings continued to rely on Tamarack through winter before switching to Black Spruce in March. Thus, even though the upright cones of Tamarack (Fig. 1.9) are likely less prone to shedding their seeds than White Spruce, Tamarack seeds are just as accessible to Redpolls, allowing them to outcompete white-wings.

This leaves one conifer: Black Spruce. The cones of Black Spruce are described as semi-serotinous, meaning that the seeds are held in the cones for extended periods of time—up to 19 years (Chai and Hansen 1952). But unlike in the serotinous-coned Rocky Mountain Lodgepole Pine *Pinus contorta* var. *latifolia*, whose cones remain closed for many years to decades, the mature cones of Black Spruce open within their first year, mostly in spring. However, many scales open only slightly (Fig. 4.1A) and the narrow gaps between the rigid scales secure many of the seeds, especially for the first two years (Viglas et al. 2013), and make them difficult to access. So difficult that I have not seen another animal besides white-wings, and on just one occasion, a Red-breasted Nuthatch *Sitta canadensis*, attempt to remove seeds from open or partially open Black Spruce cones. Red Squirrels *Tamiasciurus hudsonicus* harvest closed Black Spruce cones, but these are a seed of last resort

FIGURE 4.1 (A) Partially open Black Spruce *Picea mariana* cones and (B) White-winged Crossbills *Loxia l. leucoptera* foraging for seeds in such cones. Seeds are most likely to be retained between the least separated scales. Because the narrow gaps between the rigid scales are difficult to expand, the slender upper mandible of the White-winged Crossbill appears critical for accessing seeds. B taken by G. McElroy.

even for them (Brink and Dean 1966) and rather few are harvested (Prévost et al. 1988). Meanwhile, white-wings can remove Black Spruce seeds (Fig. 4.1B) because their slender bills (Fig. 3.6) slide between the rigid scales with relatively little resistance. Indeed, I believe the white-wing's slender upper mandible compared to those of other crossbills reflects an adaptation for feeding on seeds in slightly open Black Spruce cones. When I compared the intake rates on Black Spruce of captive white-wings to reds, the larger-billed reds were the only ones able to remove seeds from tough closed cones (albeit rather slowly). Once the cones began to open, however, the intake rates of white-wings greatly exceeded those of the reds, highlighting the advantage of the slender upper mandible (Benkman 1987b).

Although the seeds of Black Spruce are eaten much less often than those of White Spruce and Tamarack, and white-wings never have particularly high intake rates when foraging on Black Spruce, its seeds are reliably and profitably available to white-wings alone. When other seed crops fail, white-wings can survive on Black Spruce even during the short, cold days of a Canadian winter, although they may need to forage nearly continuously. Another feature of Black Spruce that makes it a reliable resource is that its annual seed crops are less variable than those of White Spruce and Tamarack (Fowells 1965). During most years in any given location, few seeds are produced by either White Spruce or Tamarack. In contrast, Black Spruce produces at least a small seed crop almost every year, and cones that retain seeds from previous years accumulate in the canopy.

I did not anticipate that the morphology of a crossbill would be specialized on a nonpreferred, seldom-used conifer. Ernst Mayr, who was the most distinguished evolutionary biologist among all ornithologists of the twentieth century, would likely not have anticipated this idea either. He wrote when discussing specialization in crossbills (Mayr 1947: 281), 'In each region a geographical race evolved which is optimally adapted for feeding on the most common conifer of the region.' In northern Europe, the three species of crossbills mentioned earlier in this chapter regularly forage on the conifers on which they appear morphologically specialized. The general expectation is that morphological or phenotypic specialization correspond to diet or ecological specialization. However, the idea that natural selection could favor phenotypic specialization on a nonpreferred resource was later given a conceptual foundation by Robinson and Wilson (1998). They named the concept 'Liem's Paradox' after Karel Liem, who had observed that some cichlid fishes infrequently fed on the resources for which they had obvious morphological adaptations. Robinson and Wilson showed that such a pattern would result when specialization provides greater access to a nonpreferred yet reliable resource (reliable especially when other resources are likely to be scarce) while not compromising the ability to use preferred resources. Some resources are plentiful and easy to use by a diversity of phenotypes (preferred resources) whereas for other, less profitable resources phenotype (morphology, physiology) matters, favoring phenotypic specialization on the less profitable but 'private' resources. This could give rise to a community of specialists, each adapted to a distinct nonpreferred or low-ranked resource, with diet convergence on preferred resources when abundant.

Although Robinson and Wilson (1998) confined their discussion to fishes and believed Liem's Paradox was less applicable to terrestrial vertebrates, I think they provide a general framework for explaining the uncoupling of morphological and diet specialization in white-wings. Morphological specialization for foraging on Black Spruce improves access for white-wings while not interfering with their ability to use the sometimes plentiful and generally preferred seeds of White Spruce and Tamarack. More recently, similar logic has been applied to a diversity of interactions ranging from feeding adaptations in snakes (Wiseman et al. 2019) to plants adapting to pollinators (Aigner 2001; Pauw et al. 2020).

Nevertheless, my earlier conclusion for specialization of white-wings on Black Spruce would have been more persuasive if I could have tested adaptation quantitatively. That is, if I could test whether white-wings match some measure of optimum bill structure alluded to by Mayr for foraging on Black Spruce. However, I didn't believe it would be feasible to predict an optimal bill size using just white-wings. The limited range of variation in bill size within white-wings didn't seem adequate. And I couldn't easily combine data from both white-wings and reds to predict an optimum because they differed both in bill size and shape. Instead, my focus shifted to the diversity of Red Crossbills in the west, where size variation was substantial and continuous with relatively little variation in shape. At about this time, Jeff Groth (Groth 1988, 1993a) was categorizing the size variation into 'call types' based on variation in their vocalizations, especially their contact calls. Groth recognized eight call types, each of whose size variation approximated that found within an average songbird species. Five call types are most common in the west, and a sixth, a mostly Mexican call type, is also found in the mountainous borderlands of New Mexico and Arizona. This presented an opportunity to test for specialization in multiple call types, which allows for a more general test of my hypothesis.

Predicting and testing specialization

I began thinking about predicting which conifers different call types might be specialized on in the west after my thoughts on white-wing specialization were published (Benkman 1987a,b). Three features seemed critical for specialization (Benkman 1993a). One was that the conifer held seeds in its cones through late winter and spring. Many North American conifers shed most to nearly all their seeds by early winter (Fowells 1965). Thus, late winter through spring is a period of diminishing cone-seed and the greatest food scarcity (Fig. 2.2). I was influenced in focusing on periods of greatest food scarcity by others working on seed-eating birds. This included Ian Newton (1967, 1972), who worked on British finches including crossbills and made a compelling case for adaptation to winter foods; my PhD advisor Ron Pulliam, who worked on North American sparrows whose conical bills could only be explained as adaptations to their nonbreeding diet of seeds rather than the insects they relied on while breeding (Pulliam 1975); and my postdoctoral advisor at the time at Princeton University, Peter Grant. Along with his wife Rosemary, colleagues and students, Peter provided now classic evidence that natural selection on bill structure was most intense during food scarcity and that the bill structure of Darwin's ground-finches *Geospiza* spp. was related to their nonbreeding season diets (Grant 1986; Grant and Grant 2014). It seemed only logical that natural selection would overwhelmingly favor crossbills that were especially well adapted to conifers that held seeds reliably during periods of greatest food scarcity. Under these conditions, the ability to feed efficiently could mean the difference between life and death.

A second critical feature was that the conifer consistently produced seeds or held seeds from year to year. As discussed in Chapter 2, seed crops vary between years and most conifers produce seed crops irregularly. Thus, conifers that produce seed crops most regularly were those that I expected to allow crossbill specialization.

A third critical feature was that the seeds during late winter and spring were encased in partly open or reclosed cones, for which crossed mandibles seem particularly effective for extracting the seeds (Chapter 1, 'Why crossed bills?'). This third feature is somewhat redundant given the first feature of holding seeds through late winter and spring. Cones that initially open only partly or those that reclose and remain reclosed for extended periods are likely to consistently hold seeds. Those cones that open more widely are likely to shed seeds more quickly and have them depleted by less specialized conifer-seed-eating animals (e.g., Redpolls).

My argument for what characterized a key resource for specialization has elements in common with the explanation John Thompson (1994: 140) developed for the evolution of extreme specialization in free-living animals. John set out four conditions favoring such specialization, which I quote below in italics followed by comments about white-wings and Black Spruce.

1. *The victim species is abundant throughout the year (or the part of the year that the forager is searching for food) or during periods of time when other potential foods are scarce.*

The victim species for crossbills is of course the conifer or more specifically the conifer's cone-seed. Black Spruce seeds are available throughout the year and especially in late winter–spring, when many of its cones open and other conifer seeds are generally scarce and declining.

2. *The victim species is predictably available, locally or regionally, year after year.*

Black Spruce is a fairly regular cone producer, and cones hold seeds for several or more years in a canopy seed bank. In addition, Black Spruce has a vast distribution, making it likely that large seed crops can be found every year somewhere.

3. *Successful capture, handling, or digestion of the victim as a major component of the diet requires specialized foraging techniques, morphology, or physiology.*

Few species besides white-wings even attempt to forage for seeds in Black Spruce cones. The crossed mandibles and specifically the slender upper mandibles appear critical in providing white-wings access to the seeds between the cones' slightly open scales.

4. *The victim moves slowly or is completely sessile, and it is easy to find.*

This applies to Black Spruce, but it applies to all conifers and thus does not help distinguish between them. John initially developed these ideas in his 1982 book. I had read this book soon after it was published, but I am uncertain of how it—in particular the sentence on page 23 where his ideas on the evolution of specialization are expressed—influenced my later thoughts on this topic. Although it might help explain why I've always felt that the argument I made was just common sense.

Based on the three general features mentioned above for crossbills, I reviewed the forestry literature to narrow down the cone-bearing conifers to those that could seemingly serve as suitable resources that would favor specialized morphologies. Especially useful was the seed fall data, measured by putting out seed traps on the ground to collect falling seed. If most seeds were collected in autumn and few if any fell later, then it was safe to assume that most seeds fell out of the cones (or were eaten) before winter. If a sizeable proportion of the seeds was collected in the spring or later, then it implied that many seeds were held in cones over winter and spring. The forestry literature, especially the reviews compiled by Fowells (1965), also provided information on the frequency of large cone crops. With such data at hand, I narrowed down the conifers to five in the Pacific Northwest, where Red Crossbill diversity and abundance were greatest in North America, and where I was hoping to conduct my research.

Western Hemlock *Tsuga heterophylla* is the smallest-seeded and -coned of what I called the five 'key' conifers (Fig. 4.2). Western Hemlock occurs in the Pacific Northwest and reliably holds seeds in its cones through spring along the rainy coast, where humid conditions reclose the cones and keep them that way for much of the year. Hemlock is also a fairly regular cone producer. Mature trees produce cones during most years, and during most years a large seed crop can be expected somewhere along the coast from south-central Alaska to northern California. The crossbill associated with Western Hemlock is the smallest of all North American Red Crossbills, categorized as Type 3 (Fig. 1.4), and corresponds to the subspecies *L. c. minor*. This subspecies was described from specimens collected in the Great Lakes region (Michigan; Payne 1987) where it occurs occasionally in large numbers (Benkman and Young 2020). Here Type 3 is often associated with Eastern Hemlock *T. canadensis*, which has cones similar to those of Western Hemlock. The occasional occurrence of large numbers of Type 3 in the Great Lakes region presumably reflects both the progressive increase in numbers that occurs following a series of years with large seed crops and the occasional large-scale cone failure causing the crossbills to erupt from the Pacific Northwest. That Type 3 rarely persist for long

FIGURE 4.2 Illustrations of the cones and seeds of five key conifers on which different call types of Red Crossbills *Loxia curvirostra* were predicted to morphologically specialize. All drawings from Sudworth (1917, 1967) except Douglas-fir *Pseudotsuga menziesii* cone from iStock.com/ibusca.

in the east presumably reflects the less humid conditions than along the coast in the Pacific Northwest.

The second predicted key conifer is Sitka Spruce *Picea sitkensis*, with a larger cone and thicker scales than Western Hemlock (Fig. 4.2). Sitka Spruce is even more closely tied to the humid northwest coast than Western Hemlock, and often holds its seeds in cones until the following summer for the same reason mentioned for hemlock. It was less clear that there was a crossbill other than Type 3 associated with this species. I thought Type 1, which is larger than Type 3, might turn out to be specialized on Sitka Spruce. A few individuals had been found in the Pacific Northwest; however, Groth had captured and described Type 1 in the southern Appalachians. Furthermore, Tom Hahn, who had spent many days along the west coast of the Olympic Peninsula of Washington capturing crossbills as a graduate student at the University of Washington, mainly encountered Type 3 and observed them foraging on both hemlock and spruce. Of course, there might not be a Sitka Spruce specialist. It was possible that Western Hemlock and Sitka Spruce were not all that distinctive to a crossbill, and that Type 3 had a bill structure equally adapted for foraging on both.

Douglas-fir *Pseudotsuga menziesii* is the third key conifer (Fig. 4.2). Two subspecies of Douglas-fir are widespread in the west. A larger-coned subspecies occurs in the Pacific Northwest (*P. m. menziesii*), and this seemed to be a more suitable conifer than the subspecies in the Rocky Mountains (*P. m. scopulorum*) because it holds its seeds more consistently through spring. Similar to Type 1, Type 4 has a larger bill than that of Type 3 and is commonly found in Douglas-fir, especially in the Pacific Northwest. Because Douglas-fir has larger, thicker cone scales than Western Hemlock, a larger bill is expected for Type 4.

Rocky Mountain Lodgepole Pine, which occurs from the Yukon south through the Rocky Mountains to southern Colorado, is the fourth key conifer (Fig. 4.2). Seed trap data reveal that seeds fall from the cones year-round, indicating that seeds are available to crossbills in partially open cones throughout the year. This is presumably related to Rocky Mountain Lodgepole Pine producing serotinous cones, which hold their seeds in hard, closed casings for decades. Over time, especially supported by subsequent and extensive observations by my students, I came to realize that what is important to crossbills relying on Lodgepole Pine are old, closed cones. As these cones weather in the canopy, seeds eventually become available to crossbills as the resin securing the scales together weakens, causing gaps to form between the scales that with effort crossbills can spread apart to reach the seeds. After the developing cones mature and dry in autumn but prior to years of weathering, closed (serotinous) cones are impenetrable to crossbills. The crossbill associated with Rocky Mountain Lodgepole Pine is the Type 5 crossbill, which has a considerably larger bill than those of Types 1 and 4.

A fifth key conifer is Ponderosa Pine *Pinus ponderosa* (Fig. 4.2). Like Douglas-fir, Ponderosa Pine is widespread in the west and has a subspecies in the northwest south into California (*P. p. ponderosa*), and a subspecies confined to the Rocky Mountain region (*P. p. scopulorum*). But unlike Douglas-fir, the subspecies in the Rocky Mountains seems more suitable for specialization; it holds seeds more consistently through spring. In addition, I later learned that the large seeds of the farther west subspecies are difficult for crossbills to husk, such that crossbills are likely limited to feeding on only those trees with smaller than average-sized seeds. The largest and most widespread call type in the west is Type 2, which is often associated with Ponderosa Pine, especially in the Rocky Mountains.

With this background, I wrote a two-page proposal while I was a postdoc at Princeton University. I was seeking additional postdoctoral funding to work with Dolph Schluter, who as a graduate student of Peter Grant worked on Darwin's finches, especially the ground-finches. Among many clever things, the two of them figured out how to characterize the interaction between bill and seed structure, such that they were able to predict the observed number of ground-finch species and their mean sizes on different islands in Galápagos (Schluter and Grant 1984). Their work was extraordinary. Few had such a thorough understanding of a community of animals while also being able to test their understanding in such a rigorous and insightful manner. I had a more modest goal of testing quantitatively whether each call type in the northwest had a bill structure that approximated the optimum for foraging on the different proposed key conifers. In addition to being the ideal sponsor and advisor, Dolph was on the faculty at the University of

British Columbia, which was an excellent base due to its proximity to a diversity of crossbills and conifers.

My goal was to capture crossbills of the five call types mentioned above, and measure the time they required to remove and husk seeds from cones of each of the five predicted key conifers. The continuous distribution of bill and body sizes among the five call types should enable me to relate these times along with estimated energy requirements to variation in bill size. I could then test whether the different call types were adapted to a key conifer by determining whether their bill size matched that predicted to be most efficient or optimal.

After receiving the fellowship, I packed my belongings and drove west from Princeton. I arrived in Vancouver at the end of May 1989. Soon after arriving, Dolph and I visited the animal care facility where he arranged for me to house the crossbills. This was adjacent to where Dolph would later have ponds built, and where he has continued to conduct extraordinary and pioneering studies with his students and colleagues on diversification and speciation in threespine stickleback fish (*Gasterosteus*).

Once I had the aviaries arranged, I set out to capture crossbills. During my first trip, I picked up Tom Hahn in Seattle and we drove to eastern Washington. Tom brought along a cage with a captive from each of three call types (2, 3 and 4) to use as decoys. Crossbills are highly social, and captives readily call upon hearing their own kind. This can cause free-flying crossbills to fly near and even land on the cage, especially if they are of the same call type as the captive. Indeed, the vigorous response to individuals of their own call type helped motivate Jeff Groth to explore the link between vocalizations and morphology in Red Crossbills. Not surprisingly, decoys are handy for luring crossbills into mist nets for capture, especially when you are unable to find a location where crossbills regularly come to the ground.

Spending time with Tom was quite helpful. I had limited experience with Red Crossbills in the northwest, whereas he had spent numerous days capturing and observing crossbills of multiple call types and was more attuned to their vocalizations than I ever will be. Tom described his observations of the general conifer associations and seed use by the different call types I described earlier, and he confirmed the extensive size overlap between Types 2 and 5. Type 2 averaged about 0.3 mm larger in bill depth, but few individuals of these two call types can be distinguished by external measurements alone. I don't think I had realized how similar they were; at the very least, I hadn't thought much about it until our discussions on the trip. It was perplexing because classic ecological theory predicts that if two species are too similar, then one of them should outcompete and replace the other. Or if they aren't reproductively isolated into separate species, they will blend from interbreeding and admixture.

Given the differences Tom described in conifer use by Types 2 and 5, I suspected they differed in some other way than simply bill size. I had recently published research discussing seed husking efficiency and palate structure (Chapter 1, 'A cost of specialization'), so an obvious explanation came to mind. As described in

Chapter 1, crossbills—like other cardueline finches—use their tongue to secure seeds in a lateral groove in the horny palate while they crack the seed coat with the especially sharp tomium on the corresponding side of their lower mandible (Fig. 4.3). Because the seeds of Ponderosa Pine are about 10 times larger than those of Lodgepole Pine (Fig. 4.2), you would expect Type 2 to have wider lateral grooves than Type 5. Individuals of each call type therefore would be more efficient at husking seeds from differing pines, causing differential pine use and enabling coexistence.

Tom and I were successful in capturing Type 2 and a handful of Type 4 crossbills. By early October, I had captured crossbills of all the call types except Type 1. In fact, I failed to even hear Type 1 crossbills. I had also tried unsuccessfully to capture Type 6 crossbills in southeastern Arizona. Type 6 (subspecies *stricklandi*) is the largest crossbill in the New World and occurs mostly in Mexico. Sometimes it occurs in southeastern Arizona, which is where I tried to find them. It was a long shot but capturing them would have extended the range of bill sizes of my

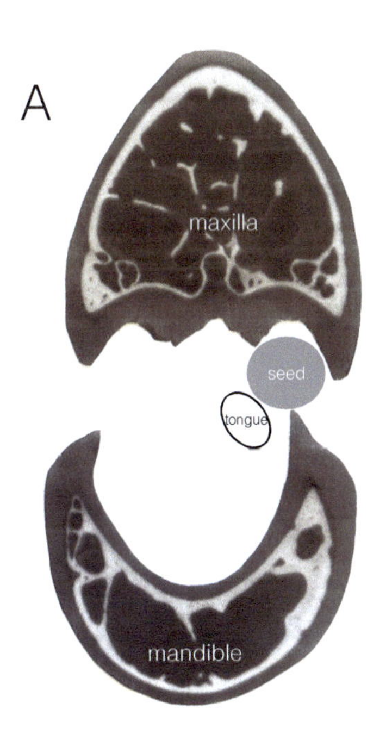

FIGURE 4.3 (A) A cross-section of the upper mandible (maxilla) and lower mandible (mandible) while a Red Crossbill *Loxia curvirostra* secures a seed in a lateral groove in the maxilla with its tongue and cracks the seed covering with the mandible. Note the especially sharp and vertically oriented mandibular tomium on the side where it cracks the seed coat. The cross-sections are from a CT scan of a Red Crossbill provided by Haley Martens. (B) A female Red Crossbill (most likely Type 2) husking a Ponderosa Pine *Pinus ponderosa* seed. The pale object in her mouth is the tongue securing the seed in the groove in the palate on the side (her left) opposite to which the lower mandible crosses (to her right). Photograph of Red Crossbill by Tony Leukering, taken in Colorado on 19 November 2004.

captives, aiding my ability to detect curvilinear relationships in the foraging data. I had also thought they might be adapted to the large, hard cones of Apache Pine (*Pinus engelmannii*), which would have provided an additional test of my hypothesis for favoring specialization.

Although I was ultimately unsuccessful in capturing Type 6 crossbills, the drive south to Arizona allowed me to find out if the palate grooves might differ in width between Types 2 and 5. Jeff Groth had finished his fieldwork for his dissertation, where he recorded the vocalizations of hundreds of Red Crossbills he had collected and prepared as scientific study skins. Although Jeff prepared few as skeletons, he did have skeletons for both Type 2 and 5. I asked if I could look at them and Jeff agreed. I drove to the Museum of Vertebrate Zoology at the University of California at Berkeley, where Jeff placed two skulls on his desk. Not knowing which was which, I turned the skulls over to look at the palates and guessed the one with a wider lateral groove was the Type 2. Surprised, Jeff asked how I could tell. This was exciting to me, although not so much to Jeff. He had already measured numerous external and skeletal features, none of which alone or together enabled him to distinguish individuals of Type 2 from Type 5. But he had not measured palates, and he had little opportunity to change that for his dissertation. It is a shame that high-quality CT scans (Fig. 4.3A) had not become widely available to ornithologists back then.

Over the summer, I started gathering thousands of serotinous Lodgepole Pine cones from recent clearcuts in the interior of British Columbia, because their hard woody cones proved especially useful for feeding crossbills and keeping their muscles and bills honed. Later I began collecting cones from the other conifers as they matured. The goal was to have sufficient numbers of cones to time each captive crossbill removing 10 seeds from each of 20 cones from each conifer, where 10 cones would represent closed late summer cones and 10 would represent the slightly open or open but reclosed cones of winter. I needed to gather many more cones because I used only those whose lengths approximated the average for the species or subspecies and that were not of unusual shape.

I finished after several months of foraging trials. Some of the data were striking. Figure 4.4A is an example. It shows the time taken by each of 27 crossbills to extract seeds from Lodgepole Pine cones in relation to bill depth (see Box 4.1 for why I used time to extract seeds and bill depth in these analyses). I attribute the limited scatter in the Lodgepole Pine data in part to my ability to control for variation among the cones that I gave crossbills. I had thousands of cones to select from, so I could standardize cone size and shape, and I could readily standardize cone stage by heating the cones to open all the scales and then adding moisture to uniformly reclose them. This is the stage that I thought most closely matched the winter cones and for which I could get repeatable foraging data. The pattern was similarly compelling for Type 3 foraging on Western Hemlock cones, for which I was also able to use a narrow range of cone sizes and shapes, and it was straightforward standardizing reclosed cones. Indeed, when Anna Lindholm conducted

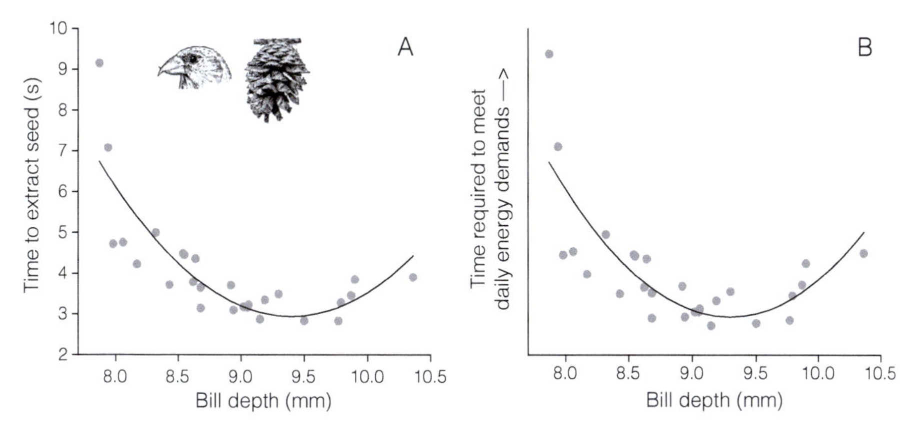

FIGURE 4.4 (A) The time to extract seeds from reclosed Rocky Mountain Lodgepole Pine *Pinus contorta* var. *latifolia* cones is curvilinearly related to bill depth, with individuals having a bill depth of 9.45 mm expected to spend the least time to extract a seed. (B) The time required to meet daily energy demands is similarly curvilinearly related to bill depth (see text). The optimal bill depth is that which minimizes the time needed to meet energy demands, which is 9.30 mm. Each data point represents the mean for one individual removing 10 seeds from each of 10 cones; the curves represent quadratic equations ($P < 0.0001$ for full models and quadratic terms). I excluded Type 5 from these analyses even though they are the call type predicted to be adapted to Lodgepole Pine because they extracted Lodgepole Pine seeds more quickly than predicted based on their bill depths. Measurements I made of palates indicate that the width of the upper mandible narrows more rapidly near the base in Type 5 than in for example Type 2. This might explain why Type 5 were even more efficient than predicted given their bill depth (Benkman 1993a). (A) is modified from Benkman (1993a), Red Crossbill head is from Benkman (1999) and the Lodgepole Pine cone is from Sudworth (1917).

similar foraging trials using Western Hemlock, as I described in Chapter 1, 'Why crossed bills?', she found the same curvilinear relationships. However, I found greater unexplained scatter in the data for Douglas-fir, and the relationship for Ponderosa Pine was not curvilinear (Benkman 1993a). I was unable to gather much data using Sitka Spruce cones because the regional seed crop was poor, and I was able to gather only a limited number of cones from a seed orchard. I also had difficulty drying and then reclosing the green Sitka Spruce cones to represent winter cones. I suspect that I gathered them too early in their development.

The relationship in Figure 4.4A allowed me to predict the bill depth for which seed extraction time is minimized—the asymptote to the curve. However, one complication is that body mass and hence daily energy requirements increase with increases in bill depth. Thus, although a larger bill (and its associated large muscles and body) might provide greater access to seeds than would a smaller bill, a larger bill comes with an associated cost of greater energy requirements, as discussed in Chapter 1, 'A cost of specialization'. To take this into account, I used the relationship between lean body mass (the mass of birds for which I did not detect fat at capture) and bill depth to predict expected body mass for

Box 4.1: Why use bill depth in analyses?

I analyze time to extract seeds from cones in relation to bill depth in Figure 4.4, because bill depth is more strongly related to extracting seeds from cones than to total time to extract and husk a seed. Bill depth is also more strongly related to time to extract seeds than is either upper or lower mandible length (Benkman 1993a). This makes sense because bill depth influences the biting forces that can be exerted by the bill (Grant 1986), which are necessary to form gaps between closed or reclosed scales. A better metric of biting force in cardueline finches is body mass, but it is correlated with both bill depth and jaw muscle mass (van der Meij and Bout 2008). Moreover, body mass varies both throughout the day and seasonally, due to for example the recency of eating and drinking and variation in fat reserves, respectively. As a result, it is measured less reliably than bill depth—hence my preference for bill depth. Bill width should perhaps be even more strongly related to the forces a crossbill can exert to spread the scales apart, but I was unable to measure bill width reliably. Those who have greater consistency in their measures of bill width of crossbills find that bill depth is correlated with bill width (P. Edelaar, pers. comm. 2024). Thus, bill depth provides a reasonable proxy for the ability to both bite between scales and spread them apart.

an individual's bill depth. I then converted body mass into the expected daily energy requirements based on allometric regressions between body mass and metabolic rate from the literature (Benkman 1993a). The end product is shown in Figure 4.4B, where I've plotted the estimated time for each crossbill to meet their energy requirements foraging on Lodgepole Pine cones in relation to bill depth. The asymptote to the quadratic regression fit to the data represents the bill depth where time to meet daily energy demands is minimized. I consider this the optimal bill depth because if less time is required to meet daily energy requirements, then more time is available to perform other activities such as attracting a mate. In addition, the more quickly daily energy demands can be met the lower the risk of starvation, especially during inclement weather and when days are short and cold and seed abundance is low.

Because the estimated optimal bill depth incorporates daily energy requirements, it is shifted to a smaller size (9.30 mm for Lodgepole Pine) relative to the bill depth minimizing time to extract seeds (9.45 mm). The extent of this shift to a smaller bill depth in general depends on the extent to which the decrease in daily energy requirement compensates for the increase in the time to extract seeds. The steeper the curve to the left of the asymptote in Figure 4.4A, the smaller the difference between the optimum and time-minimizing bill sizes. In the case of Western Hemlock, the optimal and time-minimizing bill depths were the same (rounded to the nearest 0.05 mm; Benkman 1993a). In terms of testing the hypothesis that certain cone characteristics favor specialization, the mean bill depths of Types 5 and 3 matched the predicted optima (rounding to the nearest 0.05 mm) for their hypothesized key conifers. This strongly indicates that Types 5 and 3 are adapted for feeding on Lodgepole Pine and Western Hemlock, respectively. In contrast, the predicted optimum for feeding on Douglas-fir and the mean bill depth for Type 4 differed by 0.40 mm (Benkman 1993a). Because the relationship for Ponderosa Pine was not curvilinear, I could not predict an optimal bill depth. There are several reasons why my data for Douglas-fir and Ponderosa Pine did not provide suitable tests for the optimal bill depth (Benkman 1993a). For example, I used the Ponderosa Pine subspecies in the Pacific Northwest, yet with further literature review I realized the appropriate subspecies was the one from the Rocky Mountains, which has smaller cones and seeds, and from which seeds are more easily extracted and husked (Parchman and Benkman 2008). Later we ran foraging experiments using Ponderosa Pine cones from the Rocky Mountains and found that the predicted optimum approximated the observed bill depth of Type 2 (Benkman et al. 2001).

The challenge I faced in relating palate structure to seed husking time was how to measure palate structure and in particular the width of the husking groove. I mentioned this problem to others, but no one had a good solution until one afternoon Todd Hadfield, a graduate student of Dolph Schluter's, suggested using the material dentists use to make molds of teeth. While perusing different journals in the library, Todd found a paper on the systematics of bats that used similar methods. Todd was planning to work on reproductive isolation in stickleback fish but browsed a diversity of journals, and I reaped the benefits of his broad interests. I went to the university's dental school several buildings away from my office and they kindly provided both mold material and tongue depressors. I was set.

The mold material was of two components that I mixed and formed into a small ball, which I then slightly flattened on a tongue depressor. Then for each crossbill, I opened its mandibles and slid the mold and tongue depressor between the open mandibles and held the upper mandible down against the tongue depressor for two to three minutes until the material hardened (Fig. 4.5). I then gently lifted the upper mandible away from the mold.

Once I made molds of all the crossbills, I had to decide what and how to measure. Crossbills secure seeds when husking them in the lateral groove away from the base of the bill, on the side opposite to which the lower mandible crosses (Fig. 4.3B). In Figure 4.6, this is the groove along the bottom or left side of the mandible. Not knowing exactly where the seed was secured other than it was a little forward from the base of the bill, I assumed the most likely location was where the groove was widest. This is where I cut the mold perpendicular across the long axis of the mandible. The white vertical line in Figure 4.6 approximates the location along the cut where I measured what I call the husking groove width. In retrospect, I probably should have sliced the mold perpendicular to the long

FIGURE 4.5 Pim Edelaar utilising dental material to make a palate mold of a Red Crossbill *L. c. balearica* on Mallorca, Spain in 2007.

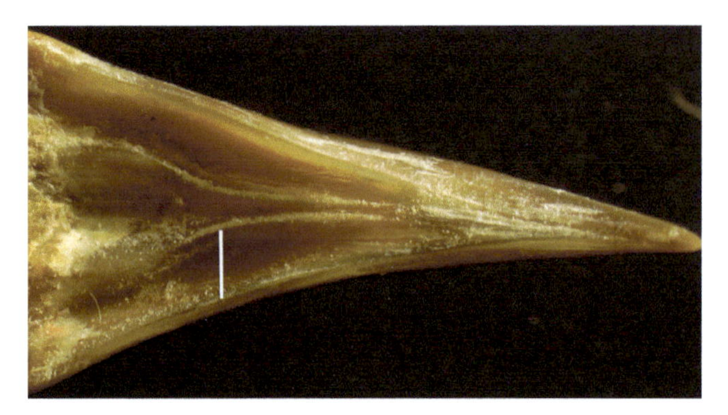

FIGURE 4.6 Photograph of the horny palate of a Red Crossbill. The white line marks where the lateral groove is widest away from the base of the mandible where I measured its width (i.e., husking groove width). This groove is on the left side of the upper mandible; the lower mandible crossed on the opposite side. Photograph taken by Zhaojie Zhang.

axis of the groove rather than of the mandible. However, given the similarity in the overall shape of the palate among crossbills, my measures should nevertheless provide a useful proxy of husking groove width.

The individual relationships between mean seed husking time and husking groove width (Fig. 4.7A) were not as striking as those for seed extraction time (Fig. 4.4A). Nevertheless, they exhibited curvilinear relationships from which I could predict the husking groove width that minimizes seed husking time. Because husking groove width was not closely linked to body mass, I assumed that the groove width that minimized husking time was the optimum. Although the match between the predicted optimal and the mean was close for all four call types, none of the individual relationships was particularly convincing. Moreover, unlike for cone size, for which I tried to use average-sized cones for each conifer, I had not ensured that the seed sizes represented the average for the conifer. This latter point was raised by a reviewer of my manuscript (Benkman 1993a),

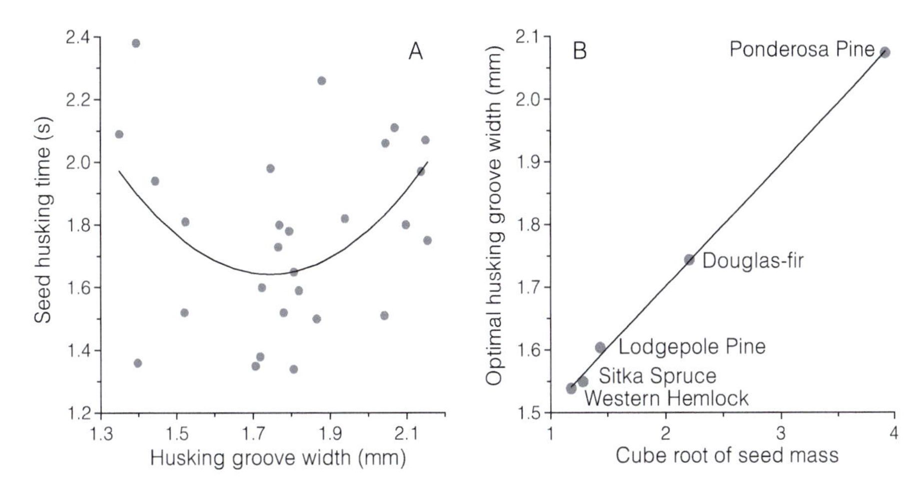

FIGURE 4.7 (A) The time husking Douglas-fir *Pseudotsuga menziesii* seeds was curvilinearly related to the husking groove widths for 27 Red Crossbills *Loxia curvirostra* of three call types. Each data point represents the mean time husking Douglas-fir seeds; the curve represents a quadratic regression. (B) The predicted optimal husking groove width was linearly related to the cube root of the mean seed mass used in the experiments. This linear relationship indicates that the optimal groove width is linearly proportional to for example the width of the seeds as they are secured in the groove. Modified from Benkman (1993a).

Diana Tomback, who works on Clark's Nutcrackers *Nucifraga columbiana* and Whitebark Pine *Pinus albicaulis*.

In an attempt to predict the optimal husking groove width for a seed of an average size for the conifer, I regressed predicted optimal husking groove width against the mean mass of the seeds used in the experiments (Fig. 4.7B). I took the cube root of seed mass to linearize it, which would better represent the width of the seed and hence its fit to my (linear) measure of groove width (Fig. 4.6). Here I include the prediction for Sitka Spruce seeds in addition to those for the other four conifers. The relationship was extremely strong ($r^2 = 0.998$, $P < 0.0001$), implying that the optimal groove width was proportional to the size of the seed, given that all the seeds are roughly of similar shape (Fig. 4.2), which made functional sense. The strength of this relationship also inspired confidence in the results. I used this relationship to predict the optimal groove width for husking an average-sized seed as found in the literature for each conifer. The match between the predicted optimum and that observed for each call type remained strong (the predicted and observed means were all within 0.05 mm of each other; Benkman 1993a).

I consider these results, especially the consistent match between the predicted optimal and that observed for seed husking groove widths, to strongly support the hypothesis that each call type is adapted to a particular conifer whose seeds are especially reliable to a crossbill (Fig. 4.2). The functional simplicity of the match between the width of the groove where seeds are secured and seed size is persuasive. Once the seed is mature, seed structure and husking time don't vary, which contrasts with the variation in accessibility of seeds throughout the year as cones mature, open and reclose, which is also more difficult to recreate.

Ten years after I published this study (Benkman 1993a), I used mark-recapture data from a banded population of crossbills in the South Hills in southern Idaho—later to be recognized as a distinct species, the Cassia Crossbill *Loxia sinesciuris*—to estimate survival in relation to bill depth (Benkman 2003). I also had intake rate data in relation to bill depth for this population (Fig. 3.7), so I was able to determine the relationship between intake rate and survival. This allowed me to convert my earlier data on intake rates into survival, which is an excellent surrogate of fitness, and thus estimate a fitness surface (Fig. 4.8; Benkman 2003). I find Figure 4.8 a useful heuristic for envisioning how different key conifers favor distinct combinations of bill depth and husking groove width that, in turn, would lead to the diversification of crossbills. The rugged surface implies that natural selection would drive populations of crossbills to the summits of the 'peaks' with individuals in the 'valleys' between them selected against, causing divergent selection favoring specialization on alternative conifers.

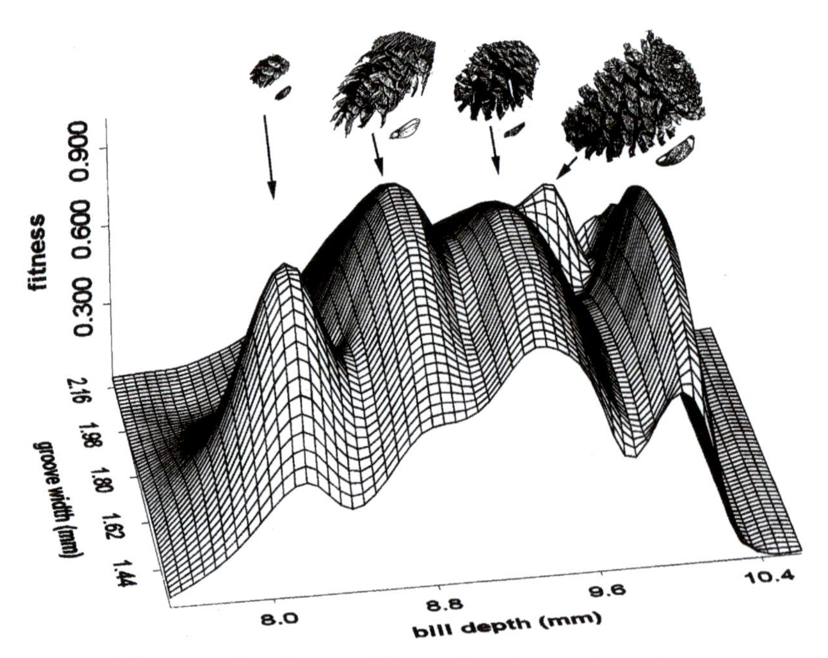

FIGURE 4.8 A fitness surface for crossbills—each point on the surface represents the estimated fitness (annual survival) for a given combination of bill depth and husking groove width—foraging on the four hypothesized key conifers: Western Hemlock *Tsuga heterophylla*, Douglas-fir *Pseudotsuga menziesii* subsp. *menziesii*, Rocky Mountain Lodgepole Pine *Pinus contorta* var. *latifolia* and Rocky Mountain Ponderosa Pine *Pinus ponderosa* subsp. *scopulorum* from left to right. The ridge along the right side represents that estimated for foraging on Lodgepole Pine in the South Hills, Idaho where the Cassia Crossbill *Loxia sinesciuris* occurs. Reproduced from Benkman (2003).

Might bill structure represent a compromise for foraging on multiple conifer species?

The generally close match between the predicted optimal bill structure for each of the four key conifers and the observed mean bill depth and especially groove width of each of the associated call types strongly suggests adaptation to a single conifer. Where there is evidence of such adaptation, the call type is perhaps best referred to as an 'ecotype', defined as 'a genetically determined phenotype of a species that is found as a local variant associated with certain ecological conditions' (Futuyma 2013). This match to one conifer is in spite of the fact that individuals of most ecotypes feed on multiple species of conifers. Indeed, in some years local populations of an ecotype will feed predominantly on an alternative conifer for several months or longer. The Lodgepole Pine ecotype is a good example. These crossbills will feed almost exclusively on Engelmann Spruce *Picea engelmannii* for much of the year when it produces a large seed crop. Engelmann Spruce cones are similar to but larger (~30% larger) than the thin-scaled White Spruce cones (Fig. 4.9) that the small, slender-billed White-winged Crossbill commonly feeds on, as discussed earlier; white-wings also breed while feeding on large Engelmann Spruce seed crops when in the Rocky Mountains. If the Lodgepole Pine ecotype's bill reflects a compromise between foraging on the thicker and hard-cone-scaled Lodgepole Pine and the thin-scaled Engelmann Spruce, then its bill depth should be smaller than that which is optimal for foraging on Lodgepole Pine. Because the average seed masses of Engelmann Spruce are so similar to those of Lodgepole Pine (3.4 mg and 3.2 mg, respectively; Benkman and Miller 1996), the optimal husking groove width is nearly identical for husking the seeds of each species (1.60 mm based on the relationship in Fig. 4.7B), so that adapting to husk one seed type would not compromise their ability to husk the other.

FIGURE 4.9 Photographs of closed cones of Engelmann Spruce *Picea engelmannii* (left) and White Spruce *Picea glauca* (right) in August when cones are nearly mature. Engelmann Spruce cones are about 30% longer and wider than those of White Spruce. The Engelmann Spruce cones were photographed in Rocky Mountain National Park, Colorado, whereas the White Spruce cones were photographed in Jasper National Park, Alberta, Canada.

Although the match between the predicted optimum bill depth for foraging on Lodgepole Pine (Fig. 4.4) and the mean for the Lodgepole Pine ecotype provides little evidence that foraging on Engelmann Spruce has led to this ecotype evolving an intermediate bill size, I further tested this with captive crossbills. I captured 10 individuals of the Lodgepole Pine ecotype and measured their intake rates on Engelmann Spruce cones (Benkman and Miller 1996). As expected, the smaller the bill, the more efficient on Engelmann Spruce. If bill size reflects a compromise between foraging on pine and spruce, then the Lodgepole Pine ecotype should be smaller than it is. This doesn't mean that the bill structures of other ecotypes do not represent compromises for feeding on multiple conifers. However, if bill structure in any of the four ecotypes discussed here reflects such a foraging compromise, I suspect it is slight from what I know about the cones and seeds that other ecotypes feed on.

Other bill functions

In recent years, some authors have stressed that because bills have other functions than just securing and handling food, bill structure potentially represents the outcome from multiple conflicting sources of natural selection (Friedman et al. 2019). The alternative function that has received the most attention is related to dissipating excess heat. Although the rhamphotheca comprising the outer surface of the bill is not vascularized, the dermis underneath is highly vascularized and thus the bill can serve as a thermal radiator (Tattersall et al. 2017). Because the dense keratin of the rhamphotheca is impervious to water, the thermal radiator function of a bill is especially important in warm environments where access to fresh water is limited and hence evapotranspiration is costly (Greenberg et al. 2012). One such environment is salt marshes, where shade and fresh water are limited. Fittingly, Song Sparrows *Melospiza melodia* of a subspecies that lives in salt marshes along the Atlantic coast have much larger bills than those of a nearby inland subspecies that has ready access to both shade and fresh water. Importantly and adaptively, especially in summer, individuals of the large-billed salt marsh Song Sparrow can dissipate about a third more heat from their bill than can individuals from the smaller-billed inland subspecies (Greenberg et al. 2012). It is quite plausible that the bill of the salt marsh subspecies is substantially larger than that which would be expected from its diet. However, it is unknown what their diet is and hence whether the large bill size might also represent an adaptation for foraging.

Crossbills occur mostly in mountainous and colder environments dominated by conifer forests, where heat dissipation from the bill is less likely to be advantageous than in Song Sparrows living in salt marshes. Instead, we might expect selection favoring a reduction in bill size to reduce heat loss (Symonds and Tattersall 2010). An advantage of the crossbill–conifer system is that we can make quantitative predictions of the optimal bill size for foraging, which enables us to judge the extent to which factors other than foraging might compromise adaptation for foraging. The one crossbill whose average bill depth is less than the predicted optimum is the Cassia Crossbill (Fig. 3.7). However, as I discuss in the next chapter, this discrepancy is likely the result of Cassia Crossbills preferentially foraging on smaller than average-sized cones. But not knowing of this cone size preference at the time, we used cones in the experiments that were representative of the average size (Benkman et al. 2001). This methodological choice likely explains why the average bill depth of the Cassia Crossbill was less than our predicted optimum. Intriguingly, no evidence was detected that thermoregulatory factors have had a widespread influence on bill size in the one group of seed-eating birds so far analyzed (Australian estrildid finches; Symonds and Tattersall 2010), perhaps because of the primacy of foraging efficiency in seed-eating birds (Tattersall et al. 2017). I think for now it is safe to assume that selection for heat dissipation or conservation has had limited effect on bill size evolution in crossbills.

What about a Sitka Spruce ecotype?

I did not find evidence that there is a Sitka Spruce ecotype, even though it seemed like a resource that could support a specialized crossbill, for the reasons explained above. I worried that neither I nor Tom Hahn had found evidence for one because such a crossbill might be rare—and rare due to excessive logging of Sitka Spruce! Consequently, I traveled to southeast Alaska in August 1992 to search for Type 1 crossbills, which, as mentioned earlier, I thought might represent a Sitka Spruce ecotype. I ended up in Juneau, where there were large Western Hemlock and Sitka Spruce seed crops. Red Crossbills were abundant, but the vast majority were the Western Hemlock ecotype. I recorded and even captured one Type 1 crossbill, but there were very few around (the remaining 33 crossbills I captured were all Type 3). In August, the crossbills foraged nearly exclusively on Western Hemlock whereas when I returned in December, they fed mostly on Sitka Spruce. Although I didn't detect any evidence that Type 1 were more restricted to foraging on spruce, I had difficulty distinguishing between the calls of the types in Alaska—I relied on Tom Hahn, who graciously confirmed the call identities of my recordings when I returned home—and thus I am not confident I could have detected a difference in conifer use even if there was one. Several years later my first graduate student, Bill Holimon, who has excellent field skills, searched widely in southeast and south-central Alaska in the summers of 1994 and 1995. Bill detected only the Western Hemlock ecotype. If there was a distinct ecotype associated with Sitka Spruce, we couldn't find it.

Over 10 years had then passed when Tom Parchman, whom I mentioned in Chapter 3, 'Cone structure', was contacted about a newly found call type. Tom had conducted a genetic study that included the different call types, which we had published the previous year (Parchman et al. 2006). He was contacted about the prospect of including a new call type in such a study. This new call type (or Type 10, as it was the tenth described call type in North America) was recently discovered by Ken Irwin, a field ornithologist who lived near the coast in Arcata, California. I called Ken to find out more about this crossbill, and what he told me suggested that he had found the Sitka Spruce ecotype that I had long predicted (Benkman 1993a). Not only did this Type 10 forage virtually exclusively on Sitka Spruce cone-seed year-round—Ken had watched these crossbills for over 10,000 hours over 10 years—but it also had a bill depth intermediate between the Western Hemlock and Douglas-fir ecotypes: 8.55 mm versus 8.20 and 8.75 mm, respectively (Irwin 2010). This intermediate bill depth was consistent with the intermediate thickness of the cone scales of Sitka Spruce, as compared to those of Western Hemlock and Douglas-fir (Fig. 4.2). And the kicker was that Ken had measured the husking groove widths of two Type 10 and found that their grooves approximated that which I had predicted in my 1993 paper based on the relationship in Figure 4.7B. It was an understatement to say that I was thrilled!

Although Ken's findings provided strong support for using conifer characteristics to predict crossbill specialization and even a new crossbill species, I thought it perplexing that Ken found the Sitka Spruce ecotype to be common only along the northern coast of California (Irwin 2010) with their occurrence decreasing northward where Sitka Spruce is more abundant. My best guess for this distribution is competition with the Western Hemlock ecotype being especially strong to the north, for two reasons. First, in Alaska Western Hemlock is much more abundant and a more prolific seed producer than Sitka Spruce, so that hemlock crossbills are often extremely abundant (e.g., Willet 1921). Second, Sitka Spruce cones to the north are smaller relative to those to the south, especially in California (Daubenmire 1968), possibly allowing the small-billed hemlock ecotype to easily switch from feeding on closed hemlock cones in the summer to Sitka Spruce cones in the autumn and winter after they open. Such a switch has been observed repeatedly by Tom Hahn in Washington and it is what I found in Alaska in 1992. Indeed, the small-billed and -bodied hemlock ecotype might be just as efficient if not more so than the larger Sitka Spruce ecotype at foraging on smaller Sitka Spruce cones once they open. This, in combination with the often large numbers of the hemlock ecotype, might prevent the Sitka Spruce ecotype from ever being common to the north. Farther south along the coast of California, the hemlock ecotype is much less abundant (Benkman and Young 2020) and here the hemlock ecotype rarely forages on Sitka Spruce (Irwin 2010). I suspect the larger Sitka Spruce cones especially in California (Daubenmire 1968) favor the larger-billed spruce ecotype over the hemlock ecotype, which could explain Ken Irwin's (2010) observations. This is speculative. However, more extensive sampling of cones than that undertaken by Daubenmire (1968), and aviary studies comparing the feeding performance of Western Hemlock and Sitka Spruce ecotypes on open and closed Sitka Spruce cones representative of those in California and farther north, would be a simple way to test this hypothesis.

Chapter 5

Coevolution: Crossbills Are More than Just Ornaments

> Interest in the coevolution of plants and herbivores has mainly centered around the food preferences of phytophagous insects and the pattern of distribution of secondary plant substances among plant species (Ehrlich and Raven 1964; Janzen 1969). The close correspondence between the phylogenetic patterns of insects and the plants they eat allows little doubt of the reciprocal effect they have on each other's evolution ... However, it is difficult to quantify the variation in secondary plant substances both within and between species and to measure their relative effect on the feeding efficiency of insects. This difficulty in measurement is largely absent in the interaction between pine squirrels (*Tamiasciurus*) and conifers. The feeding efficiency of squirrels can be closely estimated by the rate with which they ingest the female gametophyte ('endosperm') and embryos of seeds of several species of conifers. The variation in the size of seeds, the number of seeds per cone, the morphology of cones, and the seasonal availability of seeds in cones all affect the squirrels' feeding rates and can all be measured with relative ease.
>
> (C.C. Smith 1970: 349)

While contemplating studying crossbills for my dissertation, I was also revising my master's thesis for publication (Benkman et al. 1984). It focused on adaptations in Limber Pine *Pinus flexilis* that deterred its primary seed predator, the American Red Squirrel *Tamiasciurus hudsonicus*, thereby aiding seed harvest by the pine's main seed disperser, the Clark's Nutcracker *Nucifraga columbiana*. I did not consider reciprocal adaptations to Limber Pine in either the squirrels or nutcrackers, which would be a signature of coevolution. Nevertheless, I was primed for the possibility of crossbill–conifer coevolution by earlier studies on pines. An early classic on antagonistic coevolution was Chris Smith's monograph (Smith 1970) showing reciprocal adaptations between Red Squirrels and Rocky Mountain Lodgepole Pine *Pinus contorta* var. *latifolia*. Studies on nutcrackers that disperse the seeds of both Two-needle Pinyon *P. edulis* and Whitebark Pine *P. albicaulis* provided evidence of adaptation by the pines for efficient seed harvest and dispersal by the nutcrackers, and reciprocal adaptations by the nutcrackers albeit not specific to these species of pines (Vander Wall and Balda 1977; Tomback 1982).

Coevolution between crossbills and conifers seemed possible too, but none of my observations during my dissertation called out for such a process. I observed crossbills extracting and eating tens of thousands of seeds from cones, often from mere inches away. And although this allowed me to marvel at how their bill is adapted for foraging on seeds secured in conifer cones (Chapter 1, 'Getting seeds from cones'), I did not detect obvious crossbill-specific defenses in conifers that warranted further study. Features such as thick cone scales and spines could seemingly be explained as defenses directed at Red Squirrels (Smith 1970).

Even if there was coevolution between crossbills and conifers, I suspect it would have mostly added to rather than altered interpretations in my dissertation. The classic studies on seed-eating birds (e.g., Newton 1967, 1972; Pulliam 1975; Schluter and Grant 1984) indicated that considerable insight could be gained without considering whether plants were evolving defenses in response to selection exerted by birds. Moreover, the prevailing view by those studying bird–plant interactions was that strong pairwise coevolutionary interactions were rare (e.g., Wheelwright and Orians 1982; Feinsinger 1983). Indeed, just as I was starting to think about research on crossbills, a prominent evolutionary biologist questioned my suggestion of coevolution in a conversation about my work on Limber Pine. I was rightfully challenged, since demonstrating coevolution is not easy (Gomulkiewicz et al. 2007). Unfortunately, I took it at the time more as a deterrence than as an aim I should aspire to.

After finishing my dissertation, I was fortunate to receive a two-year National Science Foundation fellowship to continue research on White-winged Crossbills *Loxia l. leucoptera*. My goal was to determine how seed intake rates and social dominance influenced an individual's tendency to remain or disperse in autumn. Although I was not able to find the field conditions needed to conduct this study successfully, I ended up seeing a lot of eastern Canada. Travel, especially when it involves islands, has repeatedly proven enlightening to ecologists and evolutionary biologists. Charles Darwin and Alfred Russel Wallace come to mind. Moreover, Peter Grant was my postdoctoral supervisor at the time, and he and his wife Rosemary are rightfully famous for their research on the ecology and evolution of Darwin's finches in Galápagos (Grant 1986; Grant and Grant 2014). In retrospect, I wish I had sought island comparisons sooner but as indicated by my funded project, my focus was elsewhere.

The one island I did visit was Newfoundland, but I did not find white-wings there. This was unsurprising. White-wings can be absent from large swaths of boreal forest when few White Spruce *Picea glauca* and Tamarack *Larix laricina* cones are produced (Chapter 2, 'Crossbill responses'). However, not finding the endemic Newfoundland subspecies of the Red Crossbill *Loxia c. percna* – aka Newfoundland Crossbill – (Type 8 of Groth 1993a) was puzzling. It was supposedly a common resident, and I camped in Terra Nova National Park, which was recommended in my guidebook as one of the best locations to find Newfoundland Crossbills. But even more puzzling was that Terra Nova National Park, as well as a substantial amount of conifer forest on Newfoundland, was dominated by Black Spruce *P. mariana*. This was habitat for white-wings, the smallest-billed of all New World crossbills. Yet the Newfoundland Crossbill had the largest bill of any crossbill found regularly north of Mexico (Fig. 5.1). What was such a big-billed crossbill doing in Black Spruce forests on Newfoundland?

I had plenty of time to ponder this question as I continued my search for large developing seed crops of White Spruce or Tamarack. I left Newfoundland by ferry, eventually ending up in western Ontario for the summer. Along the drive I thought of a hypothesis for why the Newfoundland Crossbill had such a large bill. It had to do with the absence of Red Squirrels from Newfoundland. Nine thousand years ago Black Spruce colonized Newfoundland after the glaciers retreated (Jackson et al. 1997). However, Red Squirrels were unable to colonize Newfoundland on their own. In the absence of this species, I reasoned that crossbills could have

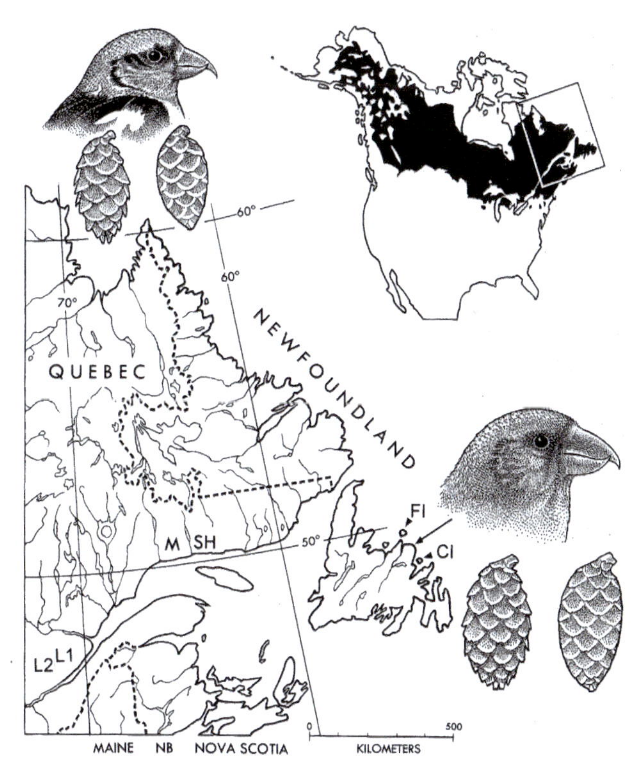

FIGURE 5.1 A White-winged Crossbill *Loxia l. leucoptera* (upper left) and a Newfoundland Crossbill *L. c. percna* (lower right), and representative partially closed and closed Black Spruce *Picea mariana* cones from the mainland (upper left) and Newfoundland (lower right). The inset in the upper right shows the distribution of Black Spruce. From Parchman and Benkman (2002). Red Squirrels occur throughout the spruce forests of the mainland and were introduced onto Newfoundland in 1963.

evolved to exploit the closed Black Spruce cones. On the mainland the closed cones are too tough for crossbills (Chapter 4), because of defenses directed at Red Squirrels. But it seemed to me reasonable that with relaxation of selection by the latter on Newfoundland, seeds in closed cones would become more accessible, opening up a resource for a large-billed Newfoundland Crossbill (Benkman 1989c). Reliance on Black Spruce is consistent with the observations made by Henrik Deichmann (pers. comm. 30 September 1988) during his surveys of birds in Newfoundland from 1957 to 1963. In an unpublished document he characterized the Newfoundland Crossbill as preferring 'the old growth black spruce forest'. The loss of squirrel defenses by Newfoundland Black Spruce could also explain why Red Squirrels had become so abundant in Newfoundland after their introduction in 1963. And the competitive effect of the now abundant Red Squirrels might explain why Newfoundland Crossbills were now so rare (Benkman 1989c) and why I never saw them. This interpretation accorded with my view of the overwhelming importance of Red Squirrels as selection agents on conifer cone structure. Coevolution between crossbills and Black Spruce initially didn't enter my mind as a contributing factor, but that eventually changed after that visit to Newfoundland.

A fortuitous conversation

After the field season in western Ontario and on my drive back to Princeton, New Jersey where I was based, I stopped at the National Museum of Canada outside of Ottawa to measure white-wings. Just before I started measuring crossbills, I fortuitously visited W. Earl Godfrey who was the emeritus curator of birds. His knowledge of the birds of Canada was immense—he literally wrote the book, *The Birds of Canada* (Godfrey 1986). I told him about my scenario for the evolution of the large-billed Newfoundland Crossbill. In response, he mentioned that I should visit the Lodgepole Pine forests atop the Cypress Hills, a sky island surrounded by prairie straddling the border of Alberta and Saskatchewan in western Canada (CH, near the top of Fig. 5.2). There, A.R. Rand and Godfrey had collected various birds including Red Crossbills in 1945 and 1948, respectively. And like the Newfoundland Crossbill, those in the Cypress Hills were common and had 'stout, blunt bills' (Godfrey 1950; Fig. 5.2), and notably Red Squirrels were absent. I spent the rest of the day measuring crossbills including those from the Cypress Hills.

Godfrey's observations took on importance several years later, after I moved to Vancouver to try to understand the ecological basis of Red Crossbill diversity, as described in the previous chapter. As during my dissertation, I read as much as I could about the cones and seeds of the relevant conifers but now in the west. As I read this literature, I was primed by the Newfoundland Crossbill to appreciate the potential importance to crossbills of geographic variation in cones and seeds. I mentioned such geographic variation between subspecies of Douglas-fir

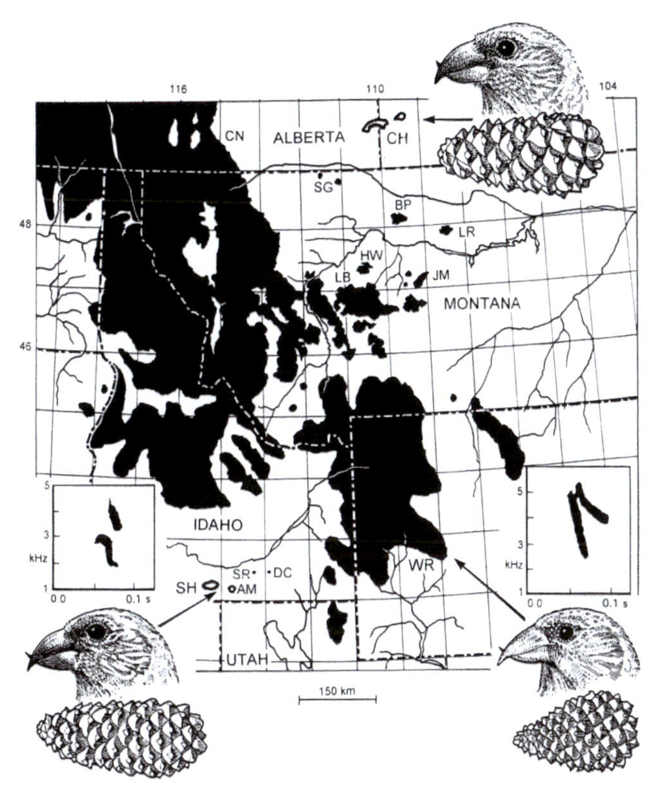

FIGURE 5.2 The distribution of Rocky Mountain Lodgepole Pine *Pinus contorta* var. *latifolia* (black) and representative crossbills and cones from the Rocky Mountains (lower right), Cypress Hills (CH; upper right), and South Hills and Albion Mountains (SH and AM; lower left). Representative spectrograms of contact calls are shown for the Cassia Crossbill *Loxia sinesciuris* (lower left) and the Rocky Mountain Lodgepole Pine ecotype (lower right) (from Benkman 1999). Red Squirrels *Tamiasciurus hudsonicus* are found throughout the range of Rocky Mountain Lodgepole Pine, except in some isolated mountains, including the Sweetgrass Hills (SG), South Hills and Albion Mountains. Red Squirrels were absent from the Cypress Hills until they were introduced in 1950.

Pseudotsuga menziesii and Ponderosa Pine *Pinus ponderosa* in Chapter 4, 'Predicting and testing specialization'. Geographic variation within Rocky Mountain Lodgepole Pine piqued my interest because of its possible relevance to Godfrey's observations. Cones of this variety varied relatively little in average size throughout the Rocky Mountains from the Yukon to Colorado but were much larger in the Cypress Hills (Wheeler and Guries 1982; Fig. 5.2) where, as I knew from my visit with Godfrey, Red Crossbills had large bills and Red Squirrels were absent.

I drove to the Cypress Hills in September 1991. Having spent a fair amount of time on Lodgepole Pine the previous two years, including gathering thousands of its cones, I could appreciate what Wheeler and Guries (1982) documented. The cones in the Cypress Hills were huge by Lodgepole Pine standards (Fig. 5.2).

Cones were longer and proportionately narrower at the base than in regions with Red Squirrels. However, what struck me most after first getting out of my car in the Cypress Hills was the abundance of Red Squirrels. They hadn't been here when Godfrey visited in 1948. They were introduced two years later in 1950 and now they were seemingly everywhere. Later I learned that their densities were about four times higher than in Lodgepole Pine forests of the Rocky Mountains where Red Squirrels are native (Benkman 1999). This further supported the hypothesis that conifers lose their defenses directed at Red Squirrels when they are absent because of relaxation of selection. Introduced Red Squirrels thrived on what to them were plentiful and relatively poorly defended seeds.

One observation made by Wheeler and Guries (1982) that puzzled me was that the pine seeds were also larger in the Cypress Hills than elsewhere. The larger seed size could not be explained by relaxation of selection from Red Squirrels because neither Smith (1970) nor his graduate student Phillip Elliott (1974, 1988a) found evidence that Red Squirrels preferred Lodgepole Pine cones with larger (or smaller) seeds. It was conceivable that abiotic conditions like aridity could favor the evolution of larger seeds in the Cypress Hills, because large seeds might aid in germination and survival (Westoby et al. 1996). However, the climate there was not particularly arid. I needed to quantify selection exerted by Red Squirrels to convince myself that selection by them hadn't favored the evolution of smaller seeds. The abundant Red Squirrels in the Cypress Hills provided an excellent opportunity for such a study.

I returned to the Cypress Hills in September 1994 to measure Red Squirrel cone preferences. Chris Milensky joined me. I mentioned Chris in Chapter 1, 'Which way to cross'. He conducted the foraging experiment testing whether seed depletion by crossbills would favor a 1:1 ratio of left to right mandible crossing directions. Because we could reach only a few cones from some trees within the forest, even with a clipper at the end of a nine-meter extension pole, we opted to search for recently fallen trees laden with many years of closed cones, of which few if any had been removed. This required a lot of walking, but once we found such a tree it was easy to remove cones. We then searched for a nearby tree from which all or nearly all cones had been cut by squirrels, with some recently cut cones remaining at the tree's base that we could gather. Red Squirrels often either mostly avoid or completely harvest the cones from individual Lodgepole Pine trees. For example, Elliott (1988a) found that Red Squirrels harvested all the cones from 62% of the 856 Lodgepole Pines on the two squirrel territories he studied in the Front Range in Colorado. This pairing of trees—one avoided to one preferred—allowed us to determine what cone characteristics were preferred and conversely those that deterred Red Squirrels. Pairing trees that were also equidistant from the squirrel's midden was important, because Red Squirrels become increasingly selective of cones with increasing distance from the midden where the squirrels' cache is located (Elliott 1988a). We were able to locate 30 such pairs of trees, from which we collected cones.

A chance comparison

While in the Cypress Hills, Chris and I climbed a fire tower from where we could see an isolated group of mountains, the Sweetgrass Hills, to the southwest in Montana (SG in Fig. 5.2). On our way back to New Mexico State University in Las Cruces, where I had started as a faculty member the previous year, we stopped in the Sweetgrass Hills and hiked to some patches of Lodgepole Pine, where we were able to use the clipper atop our extension pole to cut cones along the edge of the forest. The Sweetgrass Hills lacked Red Squirrels, and what impressed me was that the cones were small, unlike those in the Cypress Hills. The implication was that the large cones in the Cypress Hills were not simply the result of relaxation of selection by Red Squirrels, otherwise the cones in the Sweetgrass Hills would also be large. Something else was happening. We did not hear or see crossbills in the Sweetgrass Hills, but we were there for only a few hours. I reasoned that the small area of Lodgepole Pine in the Sweetgrass Hills—later I was to find out there is only about 5 km^2—might not be enough to support many crossbills year after year. Later surveys over two years by Adam Siepielski confirmed that crossbills were uncommon in the Sweetgrass Hills (Siepielski and Benkman 2005). In contrast, the Cypress Hills had about 80 km^2 of Lodgepole Pine, and Red Crossbills had been common there prior to the introduction of Red Squirrels. It was then that I realized coevolution between crossbills and conifers was a possibility!

After returning to Las Cruces, Chris measured various traits of the cones. As with previous studies (Smith 1970; Elliott 1974, 1988a), we didn't detect evidence that Red Squirrels exerted selection on Lodgepole Pine seed mass (Benkman 1999; Benkman et al. 2003). Thus, we could rule out relaxation of selection by Red Squirrels as the cause of large seeds in the Cypress Hills.

Of particular note was the shift among the different areas in the relative thickness of the scales located away from versus those near to where the cone is attached (distal and basal scales, respectively). In the Rocky Mountains, the basal scales were thicker than the distal scales (four leftmost cones in Fig. 5.3) presumably to deter Red Squirrels, which bite off successive scales starting at the base of the cone to access the distal half where most of the seeds are located. In the Sweetgrass Hills, basal scales were relatively thin and did not differ in thickness from the distal scales, consistent with a loss of defense against Red Squirrels. In contrast, the distal scales were thicker than the basal scales in the Cypress Hills (two rightmost cones in Fig. 5.3). The distal third is where crossbills usually spread apart the scales to access seeds (Fig. 1.7). That thicker scales impeded crossbills was evident from my earlier measurements showing that the time required to extract seeds from closed spruce cones increased with increasing scale thickness (Fig. 3.1). The obvious explanation therefore was that thicker distal scales were an evolutionary response to selection exerted by crossbills. Crossbills should avoid cones with thicker distal scales because they would impede crossbills from spreading apart the scales to reach the seeds at the base (Fig. 1.7B).

FIGURE 5.3 Serotinous Rocky Mountain Lodgepole Pine *Pinus contorta* var. *latifolia* cones with their basal ends (where they attach to the branch) toward bottom of photograph. The four cones on the left are from four trees in the Laramie Range, Wyoming, and illustrate the variation in cone structure among trees and the relatively large basal scales and wide cone bases characteristic of cones from areas with Red Squirrels *Tamiasciurus hudsonicus*. The two cones on the right are from one tree in the South Hills, Idaho, and illustrate the relatively large and thick scales at the distal end of the cone characteristic of areas where the pine has evolved in the absence of Red Squirrels and where crossbills occurred in high densities, such as the Cypress and South Hills.

In support of these interpretations, Red Squirrels required 37% less time to remove a seed from cones in the Cypress Hills than the time Phil Elliott recorded and kindly sent me for Red Squirrels in the Colorado Rockies (5.8 vs. 9.3 seconds per seed, respectively). Conversely, seven captive crossbills timed foraging on Lodgepole Pine cones required 67% more time to extract a seed from Cypress Hills cones than from those from the Rockies (7.0 vs. 4.2 seconds per seed, respectively), consistent with elevated crossbill defenses.

Finding another Cypress Hills

Coevolution occurs between two species when they both evolve in response to the natural selection they exert on each other. In the case of the coevolutionary arms race that I envisioned between crossbills and Lodgepole Pine, seed predation by crossbills favored the evolution of trees with elevated seed defenses (thicker distal scales) because those trees were avoided and more likely to leave offspring, causing an increase in distal scale thickness in successive pine generations. This increase in scale thickness in turn should favor crossbills with deeper bills to increase seed access. The inference that the large bills of the crossbills collected in the Cypress Hills in the 1940s and the thick-distal-scaled cones represent genetically based evolutionary responses is reasonable, given that both the cone traits of multiple species of conifers and the sizes of the bills of birds have been found to have high heritabilities (conifers: Khalil 1984; Singh and Chaudhary 1993; Kaya and Temerit 1994; Matziris 1998; birds: Boag and van Noordwijk 1987; Grant and Grant 1989). Chris Smith had also told me at the time that he had evidence that many Lodgepole Pine cone traits were heritable, but this remains unpublished. Evidence for high heritability of bill depth in crossbills was published later (Summers et al. 2007).

Local differences in traits can also arise because of phenotypic plasticity in direct response to the environment. For example, herbivory can induce plant defenses that reduce further herbivory. However, it is unlikely that phenotypic plasticity in response to seed predation accounts for the abovementioned differences in cone traits. If the differences in cone traits among trees were the result of such induction rather than an evolutionary response, then trees whose cones had been harvested by Red Squirrels year after year should have responded with elevated defenses, whereas those that have not been harvested should lessen defenses. We found no such evidence of phenotypic plasticity: trees repeatedly harvested by Red Squirrels year after year continued to produce poorly defended cones, while those avoided continued to produce well-defended cones.

Regrettably, in the absence of a common resident crossbill, there was little opportunity to measure either selection exerted by crossbills on Lodgepole Pine or reciprocal selection on the crossbills. In the absence of such a crossbill, I initially sought additional isolated ranges of Lodgepole Pine to determine if the differences between the Sweetgrass Hills and the Cypress Hills were repeated. Documenting whether patterns are repeated is a strategy often employed by biologists to help eliminate the possibility that the pattern was due to chance.

I reviewed distribution maps of Lodgepole Pine (Fig. 5.4) and of Red Squirrels, trying to determine where I could find replicates of isolated ranges containing Lodgepole Pine but without Red Squirrels. I could find several small ranges without Red Squirrels, but I could not find ranges without them that were comparable in size to the Cypress Hills, which might support a resident crossbill population. I remember one afternoon in 1996, examining maps in an attempt to find additional replicate ranges in response to reviews of the grant proposal I was revising. I had focused on ranges east of the Rocky Mountains in Montana in the previous proposal, but with Bill Boecklen's suggestion I brought some maps over to his office, which was next to mine. I think this caused me to begin investigating ranges to the west of the Rockies. However, as can be seen in the map of Lodgepole

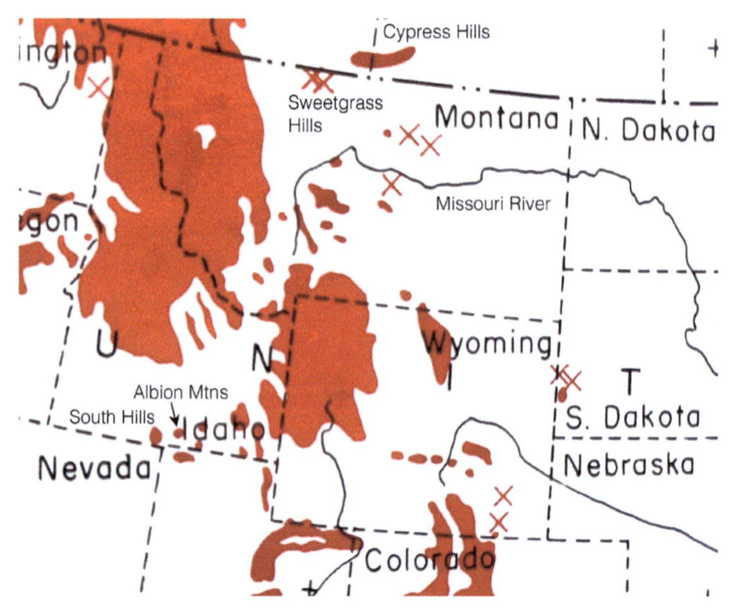

FIGURE 5.4 Map of geographic distribution of Rocky Mountain Lodgepole Pine *Pinus contorta* var. *latifolia* from Critchfield and Little (1966), which I used to locate isolated ranges with Lodgepole Pine. I have added names to some of the mountain ranges and to the Missouri River. Red Squirrels *Tamiasciurus hudsonicus* were absent from the Cypress Hills, and remain absent from the Sweetgrass Hills and the other isolated ranges in Montana north of the Missouri River, which flows east across the eastern half of the state. The isolated ranges within Wyoming all have Red Squirrels. West of the Rocky Mountains, the South Hills lack Red Squirrels, and I later discovered that the Albion Mountains lack them too. But the ranges south and east of the Albions have Red Squirrels.

Pine (Fig. 5.4), the one range I was certain lacked Red Squirrels, the South Hills (the range farthest west along the southern border of Idaho), appeared to have relatively little Lodgepole Pine. I resubmitted my grant proposal in June 1996 with additional ranges including the South Hills but without including replicates for a large range without Red Squirrels.

Later that summer in 1996, I flew to Salt Lake City and rented a car to drive to an ornithological conference in Boise, Idaho. By the afternoon I reached a Lodgepole Pine forest in the Albion Mountains, just east of the South Hills. I was elated to find that crossbills were common and Red Squirrels were absent. Moreover, the cones were like those in the Cypress Hills. I did not have a nine-meter extension pole to cut cones, so I searched for fallen branches whose cones I could collect. After collecting at least three cones from each of over 20 trees, I drove the next day to the South Hills (Fig. 5.5) where I collected more cones. Just as in the Albions, Red Crossbills were common, Red Squirrels were absent and the cones were like those in the Cypress Hills. It also appeared that the area of Lodgepole Pine was more extensive than depicted on the map (Fig. 5.4). It is no understatement that I was

FIGURE 5.5 The view south from Monument Peak in the South Hills, Idaho on 11 July 2011. Rocky Mountain Lodgepole Pine (forest stands on the hills in partial sunlight) and Subalpine Fir *Abies lasiocarpa* (forests in the shade in the near distance) are patchily distributed in a matrix of sagebrush steppe. Expansive areas of sagebrush between ranges including the Great Basin Desert, visible between the mountain ranges in this photograph, impede dispersal and colonization by American Red Squirrels *Tamiasciurus hudsonicus*.

euphoric, because it seemed that the combined pine forests of the South Hills and Albions might support a resident crossbill population that was coevolving with Lodgepole Pine. After a restless night, I drove to Boise and had lunch with Bud Tordoff. As I mentioned in Chapter 1, 'Which way to cross', Bud had conducted research on captive crossbills and knew them well. I must have seemed brash indeed when I told him, 'I think I discovered a new species of crossbill'!

When I returned to Las Cruces, I contacted the US Forest Service to find out more about the South Hills. Nearly all Lodgepole Pine in the South Hills and Albions occur within the Sawtooth National Forest. Sure enough, there was a lot more Lodgepole Pine than depicted in Figure 5.4. The total amount in the South Hills and Albions combined was nearly 70 km^2, or not much less than the estimated 80 km^2 in the Cypress Hills. The crossbills in the South Hills and Albions sounded a bit different to me, but, as I mentioned earlier, I do not profess to have the most discriminating ears. I needed to return to the South Hills to capture and measure crossbills and record their contact calls. Unfortunately, I had to wait until the following summer because classes were about to start.

I returned to the South Hills in July 1997. By then I had learned from Jack Trotter, who conducted an annual Breeding Bird Survey in the South Hills, that crossbills were common there year after year. Jack also suggested a site to net crossbills where they came for water. Based on the crossbills I measured from the Cypress Hills (two females and eight males; the mean bill depth for the two sexes was 10.00 mm) and the similarity in cones between the Cypress Hills and both the South Hills and Albions, I expected the crossbills in the South Hills to have an average bill depth of 10 mm. That was only slightly larger than the average—9.9 mm—for the adult crossbills I captured in the South Hills in 1997. I strongly suspected that their contact calls were distinct, too (Fig. 5.2).

On my last evening in the South Hills, I camped at Bostetter, which is nearly seven miles (11 km) southeast from where I was netting crossbills. Bostetter is in a valley with meadows surrounded by extensive stands of Lodgepole Pine and with a stream where crossbills drink. It became my favorite place in the South Hills. I had it to myself, and that evening offered one of my most magical, ineffable moments amongst crossbills. I was convinced that not only had I found a system where coevolution was occurring, but the coevolutionary arms race was driving this local crossbill—which I initially called the South Hills Crossbill, later to be formally recognized as the Cassia Crossbill *Loxia sinesciuris* 'the crossbill without squirrels'—to diverge and speciate. Moreover, about 30 Cassia Crossbills were flying among the pines and descending to the stream to drink. These crossbills owed their abundance to the absence of Red Squirrels, and I was able to revel in their presence. This is something I had dreamed about but would never experience in either the Cypress Hills or Newfoundland, because I failed to visit those areas before Red Squirrels were introduced and drove any locally evolved crossbills to or near extinction. On my drive home to Las Cruces, I envisaged numerous research opportunities provided by the Cassia Crossbill. Later that fall, I received funding for the grant I had submitted in June.

Finding the Cassia Crossbill came at an opportune time. John Thompson, who provided a broader context for my work on the evolution of specialization in crossbills (Chapter 4, 'Predicting and testing specialization'), was also developing the conceptual foundation for the causes and consequences of geographic variation in coevolving interactions, which he termed the 'geographic mosaic theory of coevolution' (Thompson 1999, 2005). This recognized that geographic heterogeneity in coevolution countered the assumptions of some that coevolution needed to occur uniformly range-wide, while showing why such geographic variation is widespread and important to the coevolutionary process. Moreover, John invited me to participate in the 1998 American Society of Naturalists Vice-Presidential Symposium focusing on the geographic mosaic theory. Our work fit nicely into this broader framework. Luckily, I had also hired Bill Holimon after he finished his graduate work (described in Chapter 4, 'What about a Sitka Spruce ecotype?'). Bill required minimal guidance and could be relied on to gather whatever bird or cone data was needed just in time to complete a manuscript (Benkman 1999) and for me to give a presentation in the symposium. John had provided quite a springboard for our research.

The geographic mosaic theory of coevolution posits that there are hotspots where coevolutionary interactions are strong, and coldspots where coevolutionary interactions are weak or absent. And of course, there can be all levels of variation in between. Movement and gene flow between hotspots and coldspots in turn can alter the dynamics and cause mismatches between interacting populations. Although crossbills and conifers do not lend themselves to elucidating the varied dynamics of geographic mosaics (lab studies with micro-organisms are better—e.g., Forde et al. 2004; Morgan et al. 2005), crossbills and conifers are well suited for illuminating coevolution, and the ecology of hotspots and coldspots.

Lodgepole Pine cone-seed as a resource for crossbills and the effect of Red Squirrels

Whether in the Lodgepole Pine forests of the South Hills or the Rockies, crossbills rely mostly on seeds in older, closed serotinous cones. After serotinous cones mature and dry in autumn, they are impenetrable to crossbills. But after multiple years of weathering, the resinous bonds weaken and gaps begin to form between the scales (Fig. 5.6A; Elliott 1988b), and crossbills can gain access to seeds (Fig. 5.6B). Chris Smith noted the use of such seeds by crossbills in the Front Range of Colorado in a letter to me in July 1983, long before I or my students observed such behavior. Chris wrote: 'they [Red Crossbills] even work on the lodgepole pines that have

FIGURE 5.6 (A) The distal end of a Rocky Mountain Lodgepole Pine *Pinus contorta* var. *latifolia* cone with gaps forming between the scales, some of which (to the right and lower down in front) were spread apart perhaps by crossbills. (B) Female Cassia Crossbill *Loxia sinesciuris* foraging for seeds on a weathered, more than 10-year-old Lodgepole Pine cone. Photographs were taken in the South Hills, Idaho by (A) the author and (B) Jimena Golcher-Benavides.

weathered enough to start the cones opening'. However, the availability of such seeds to crossbills is much reduced by Red Squirrels, for two reasons. The first is through their evolutionary effect on the frequency of serotiny in Lodgepole Pine populations (Box 5.1). The second is through their preemptive harvest of massive numbers of serotinous cones before they weather enough for crossbills to access seeds. To address the evolutionary effect, we need to consider the advantages of serotiny, and then the evidence that Red Squirrels have influenced the occurrence of serotiny.

Rocky Mountain Lodgepole Pine generally occurs where high-severity, stand-replacing fires are likely within a tree's expected lifespan (Fig. 5.7A). Such conditions favor pines that retain seeds in the canopy that can be released following a fire, when resources and conditions for seed and seedling survival and for seedling growth are highly favorable (Fig. 5.7C; Lamont et al. 1991; Lamont and Enright 2000). Like many other woody plants occurring in such situations (Lamont et al. 1991), Lodgepole Pine has evolved a closed-cone habit termed 'serotiny' where its seeds remain within closed woody cones that are attached securely to branches for multiple years and even decades until heated by fire (Fig. 5.7B; Lotan 1975). However, seeds retained in cones are an abundant, easily located resource for seed predators, which in turn favors the evolution of elevated seed defenses (Smith 1970). Indeed, this explains why seeds represent only about 1% of the total dry mass of Lodgepole Pine cones, most of the rest being devoted to thick woody scales that deter squirrels (Smith 1970; Elliott 1988a; Benkman 1999; Benkman et al. 2001). Although these elevated defenses greatly reduce predispersal seed predation by insects (less than 0.1% of the seeds in cones are destroyed by insects; Miller 1986), Red Squirrels still harvest large fractions of the cones (e.g., Smith 1970; Elliott 1974, 1988a). This predation reduces the benefit of seed retention and thereby acts to select against serotiny (Box 5.1; Enright et al. 1998; Talluto and Benkman 2014).

Box 5.1: Causes and consequences of variation in frequency of serotiny

Traditionally and for good reason, forest ecologists have focused on variation in the frequency of fire to account for variation in the occurrence of serotiny. More frequent (stand-replacing) fires favor higher frequencies of serotiny (Enright et al. 1998), and observations, including those contrasting Lodgepole Pine forests at low and high elevations in Yellowstone National Park, support such a relationship (Schoennagel et al. 2003). Nevertheless, there is considerable variation in the frequency of serotiny in forest stands within both the low and high elevations in Yellowstone; mature Lodgepole Pine trees produce either serotinous or non-serotinous cones (Fig. 5.7B,D), or much less often a mix (Critchfield 1980). This is where Red Squirrels come in. Theoretically, a seed predator selects against serotiny if it preferentially preys on seeds in serotinous cones, thereby depleting the canopy seed bank (Enright et al. 1998). American Red Squirrels are such a predator. Both females and males defend separate territories year-round where they harvest and cache thousands of closed cones (Fig. 5.8; Smith 1968, 1970). Because serotinous cones remain closed and hold their seeds for one or more decades, their seeds remain vulnerable to squirrel harvest for years (Elliott 1988b). In contrast, squirrels harvest non-serotinous cones almost exclusively during the several weeks between seed maturation, and scale opening and seed shedding in autumn. Consequently, the overall probability of a squirrel harvesting a given serotinous cone is about 100 times higher than that for a non-serotinous cone (Talluto and Benkman 2014). Depending on the density of Red Squirrels, this can result in extremely strong selection against serotiny.

Considering the life history of Lodgepole Pine, differential seed predation and fire frequency, one of my graduate students, Matt Talluto, showed that when Red Squirrel densities exceed one and a half squirrels per hectare, selection by squirrels against serotiny overwhelms countering selection by fire favoring serotiny, and thus nearly eliminates the occurrence of serotiny. At lower squirrel densities, the balance between selection by squirrels and fire sets an intermediate frequency of serotiny (Talluto and Benkman 2014). Matt's model accounts quite well for the variation in frequency of serotiny across Yellowstone. Moreover, Matt's model explains why the frequency of serotiny in ranges east and west of the Rockies, which squirrels were unable to colonize, is around 90% and exceeds the frequency found in areas with squirrels.

What is fascinating, beyond just the potential effects on the resource base for crossbills (fewer squirrels, more serotiny and more cone-seed for crossbills), is that the frequency of serotiny in the pre-fire forest is the best predictor of seedling density after a fire (Turner et al. 2003). When the frequency of serotiny is high, large numbers of seeds are released from serotinous cones after fire sweeps through, resulting in large numbers of seedlings. Lower frequencies of serotiny result in fewer seeds available after a fire and many fewer seedlings. The density of seedlings per hectare ranged from 600 when less than 1% of the trees were serotinous, to 211,000 when 65% of the trees were serotinous. This variation resulted in dramatic differences in nutrient flows and how plant communities developed after a fire, ranging from sparse pine seedlings and domination by various grasses and forbs, to dense growth of pines with little else. Other components such as the various animal communities (e.g., birds, mammals and invertebrate pollinators) must also vary accordingly ultimately because of selection exerted by squirrels.

FIGURE 5.7 Stand-replacing fires, like the Badger Fire in the South Hills, Idaho in September 2020 (A), are an expected disturbance for Rocky Mountain Lodgepole Pine *Pinus contorta* var. *latifolia*. Serotinous Lodgepole Pine cones (B) accumulate in the canopy for decades and open due to the heat of fire, releasing their seeds which (C) germinate the following spring and summer (seedlings in June 2021, nine months after the Badger Fire). Some trees produce non-serotinous cones (D) that open in autumn several weeks after the seeds mature. The open cones in (D) opened the previous autumn, and the closed cone on the left would have opened within several weeks after the photograph was taken.

Unsurprisingly, the occurrence and local density of Red Squirrels has a large impact on the frequency of serotiny (Box 5.1). For example, about 90% of the trees produce serotinous cones (and 10% produce non-serotinous cones; Fig. 5.7B,D) where Lodgepole Pine evolved in the absence of Red Squirrels (e.g., Sweetgrass and Cypress Hills to the east of the Rockies, and the South Hills and Albions to the west of the Rockies). In contrast, an average of 34% of the trees produce serotinous cones where Red Squirrels reside (34% is the median for 341 locations sampled by Lotan 1975; Benkman and Siepielski 2004). The local frequency of serotiny also varies inversely with the density of Red Squirrels within extensive areas of Lodgepole Pine (Box 5.1; Talluto and Benkman 2013, 2014). Thus, with increasing densities of Red Squirrels, the occurrence of serotinous cones and the size of the canopy seed bank decreases. This combined with the extensive harvest of serotinous cones by Red Squirrels reduces the availability of seeds to crossbills in Lodgepole Pine forests.

Comparisons of crossbill densities in areas with and without Red Squirrels (e.g., Rockies versus South Hills) provide an estimate of the impact of Red Squirrels on crossbills (Benkman 1999; Siepielski and Benkman 2005). Our first comparison between the Rockies and South Hills used an especially large sample of bird surveys in Lodgepole Pine from the Northern Rockies, provided to me by Dick Hutto. Crossbills were 22 times more abundant in the absence of Red Squirrels (Benkman 1999). We can also estimate the overall impact of Red Squirrels via their effect on the abundance of seeds in the canopy seed bank by comparing the densities of Lodgepole Pine seedlings after fires (Fig. 5.7C) between areas with and without Red Squirrels. This assumes that the density of Lodgepole Pine seedlings after a fire provides an index of the abundance of seeds in closed cones in the canopy, which seems reasonable (Box 5.1). This assumption is supported by the strong positive relationship between the density of closed cones (per ha) in the pre-fire canopy and the corresponding density of seedlings post-fire in serotinous Knobcone Pine *Pinus attenuata* woodlands (Agne et al. 2022; Marlin et al. 2024). The assumption is that the more closed cones with seeds, the more seeds that will be available to crossbills in weathered cones.

As mentioned earlier, an average of 34% of the trees are serotinous in areas with Red Squirrels. To estimate the corresponding density of seedlings, I used seedling densities following fires in Yellowstone National Park from Turner and colleagues (2003). I took the average of two seedling densities—one from when there was less than 1% serotiny and the other when there was 65% serotiny. The number of seedlings per hectare was 600 and 211,000, respectively, or a mean of 105,800 seedlings at around 34% serotiny. Similar results are found if I use the density of seedlings at a site with 10% serotiny instead of under 1% to estimate an average for approximately 34% serotiny. Before Red Squirrels were introduced into the Cypress Hills, which has a similar frequency of serotiny as the South Hills (92% versus 90%, respectively; Benkman and Siepielski 2004), 2.5 million seedlings per hectare were recorded following a fire (Newsome and Dix 1968). This density of seedlings is nearly 24 times greater than that estimated for an average area with Red Squirrels. The expectation therefore is that 24 times more crossbills can be supported per unit area in the absence of compared to the presence of Red Squirrels. This expectation is quite close to the 22 times greater estimated density of crossbills in the South Hills than in the Rockies (Benkman 1999). Overall, these results give credence to the hypothesis that Red Squirrels greatly suppress crossbill numbers, both by exerting selection that reduces the frequency of serotiny and by preemptively harvesting serotinous cones before crossbills can access the seeds. This further reinforces the massive potential impact Red Squirrels can have on the development of forests and the associated plant and animal communities following stand-replacing fires (Box 5.1).

A geographic mosaic of coevolution

As a consequence of their superior competitive ability, Red Squirrels drive the evolution of Lodgepole Pine cone structure throughout their vast range of overlap

(Smith 1970; Benkman 1999; Benkman et al. 2001, 2003), whereas crossbills are the main selection agent in the few ranges without Red Squirrels (Fig. 5.2; Benkman et al. 2003, 2013). It is in the largest isolated ranges where crossbills have their highest densities and exert the most intense selection (Siepielski and Benkman 2005; Benkman et al. 2013). The largest isolated ranges are also where coevolutionary arms races between crossbills and Lodgepole Pine are most evident. This geographic mosaic where both selection and the extent of coevolution varies geographically is seemingly prevalent across the diversity of life (Thompson 1994, 2005). However, the causes of such geographic structure are rarely so obvious.

Focusing first on areas where Red Squirrels overlap in distribution with Lodgepole Pine, selection by Red Squirrels appears to account for the more pronounced features of Lodgepole Pine cones (Fig. 5.3; Smith 1970; Elliott 1974, 1988a; Benkman 1999; Benkman et al. 2001, 2003; Edelaar and Benkman 2006). Red Squirrels preferentially harvest narrower cones because they are easier to both remove from branches—cones are flush to the branch and their wide, hard base makes it difficult for squirrels to bite cones off (Fig. 5.7B)—and bite through to access seeds (Smith 1970; Elliott 1974, 1988a). Red Squirrels also preferentially harvest cones with more seeds and more seed mass per unit mass of cone, because this maximizes both the amount of food energy carried to and stored in their central midden (cones are carried individually; Fig. 5.8B) and how fast seeds can be removed from the cone and eaten (Smith 1970; Elliott 1988a). The result is that in ranges historically containing Red Squirrels, cones are relatively wide with enlarged basal scales (Fig. 5.3) and have few (15–20) full seeds and a low percentage of the cone mass that is seed (~1%; Benkman et al. 2001). Chris

FIGURE 5.8 (A) Red Squirrels bite closed Rocky Mountain Lodgepole Pine cones off branches and then while manipulating the cones with their forefeet, bite off successive scales starting at the base of the cone to access the underlying seeds. Because the basal scales lack seeds, a third or more of the scales need to be removed before squirrels reach seeds. (B) Closed cones are stored in middens that are usually in the shade of a large tree near the center of the Red Squirrel's territory. Here Lodgepole Pine cones have accumulated during the fall harvest and then been buried underground; such cones are unavailable to crossbills. Photographs were taken in Yellowstone National Park (A) and Laramie Range (B), Wyoming.

Smith (1970) makes a strong case that Red Squirrels have evolved enlarged jaw muscles and associated skull features for exerting strong forces to bite through the hard, woody cone scales (Fig. 5.8A) that have evolved in response to selection by them. And thus, we infer that the reciprocal adaptations of Lodgepole Pine and Red Squirrels are the outcome of coevolution.

Although crossbills likely exert selection on the cones in areas with Red Squirrels, it is undoubtedly much weaker selection than that exerted by the latter (Box 5.2; the proportion of the seed crop harvested by Red Squirrels is probably about 100 times greater than that eaten by crossbills) and is largely limited to the subset of cones left unharvested by Red Squirrels. We have not attempted to quantify selection exerted by crossbills where Red Squirrels are present, because weak selection is difficult to detect without a massive sample size. Moreover, we have no method to assess if selection exerted by crossbills in the presence of Red Squirrels has had an evolutionary effect. To infer an evolutionary effect, we would need to compare ranges with and without crossbills but all containing Red Squirrels. Yet there simply aren't any ranges with a long evolutionary history of containing Red Squirrels but no crossbills. If Red Squirrels can colonize and survive, so too can crossbills.

As mentioned earlier, crossbill population densities are much greater in isolated mountain ranges that Red Squirrels were unable to colonize naturally than in areas with Red Squirrels. Moreover, crossbill densities increase with forest area (Siepielski and Benkman 2005): the increase in densities with area is linear, based on recent estimates of forest area. Thus, crossbill densities are especially high in larger isolated forests like those of the South Hills/Albions. This results in a proportionate increase in seed predation by crossbills (Box 5.2), which causes an increase in selection exerted on cone traits by crossbills (Fig. 5.9). This is an example of how natural selection exerted by an antagonist (e.g., seed predator) increases with concurrent increases in the occurrence of the interaction, assuming that feeding preferences are consistent (Benkman 2013). Such consistency in preferences arises presumably because crossbill foraging behavior is stereotypic (Chapter 1) and crossbills invariably seek the most profitable seeds (Chapters 2 and 3). Profitable here means the greatest caloric return for the least effort, and indeed, crossbills prefer to forage on smaller cones with thinner scales—whether they are Cassia Crossbills foraging on Lodgepole Pine, or Red Crossbills foraging on five species of pine in Europe and North America (Fig. 5.9; Benkman and Mezquida 2015).

These cone preferences by crossbills ultimately cause the evolution of larger cones with thicker distal scales, and account for the similarly thick distal scales in the South Hills/Albions and Cypress Hills (2.90 mm and 2.91 mm, respectively) relative to those in the Rockies (the mean of four areas was 2.59 mm; Benkman et al. 2001; Figs 5.2 and 5.3). In addition, we found that crossbills exert selection favoring the evolution of larger seeds (Fig. 5.9), which helps account for the large seeds in the Cypress Hills that initially puzzled me. Such selection implies that crossbills are more likely to forage on cones with small than large seeds. However,

Box 5.2: Estimating the proportion of seeds eaten

We do not have direct measures of the proportion of the Lodgepole Pine seed crop eaten by Red Crossbills in areas with Red Squirrels, but we can estimate this proportion based on measures of crossbill abundance and crossbill seed predation in the South Hills. We can also use these values to estimate the proportion of the seed crop eaten by Red Crossbills in the Rocky Mountains relative to that harvested by Red Squirrels. To do so, we first estimate that crossbills are about 20 times more abundant in Lodgepole Pine forests in the South Hills in the absence of Red Squirrels than they are in the Rocky Mountains (this chapter). Thus, we expect that crossbills eat only about one-twentieth the amount of Lodgepole Pine seeds in areas with Red Squirrels in comparison to the South Hills without Red Squirrels. Given that we found that crossbills in the South Hills consumed about 6% of the Lodgepole Pine seeds (Benkman et al. 2013), Red Crossbills in the Rocky Mountains likely consume only one-twentieth of 6%, or 0.3% of the Lodgepole Pine seeds.

Compared to estimates of cone harvest by Red Squirrels in Lodgepole Pine forests, 0.3% is a miniscule amount. Elliott (1988a) found that Red Squirrels harvested all the cones from 62% of the 856 Lodgepole Pines on the two squirrel territories he studied in the Front Range in Colorado. Red Squirrel territories often include all or nearly all the Lodgepole Pine in an area (Smith 1981) but, where pines are sparse, fewer of them occur within squirrel territories (Gurnell 1984). Thus, the overall percentage of the seed crop harvested by Red Squirrels might be less than 62%. On the other hand, 62% might underestimate the proportion of the seed crop harvested within territories, because the squirrels undoubtedly incompletely harvest cones from some trees (e.g., the remaining 38% of the trees in Elliott's study). If we assume that Red Squirrels harvest 62% of the cones in the Rocky Mountains, then they harvest about 200 times more of the seed crop than do crossbills (62% vs. 0.3%). However, we are probably underestimating the number of seeds eaten by crossbills, because our estimates are for one- to ten-year-old cones only, despite the fact that crossbills might eat a larger proportion of the seeds in cones over 10 years old. If we doubled the proportion of seeds eaten by crossbills, then Red Squirrels would still harvest 100 times more seeds than do crossbills. I suspect that this latter value of 100 times is still conservative, but nevertheless it helps account for why Red Squirrels are such an overwhelming agent of selection on Lodgepole Pine. The hundred-fold greater seed harvest by Red Squirrels than by crossbills is presumably related (at least in part) to the seven times greater body mass of Red Squirrels (227 gm; Elliott 1988b) compared to crossbills (32 gm), which provides greater power to the squirrels to harvest whole cones and access seeds in tougher cones.

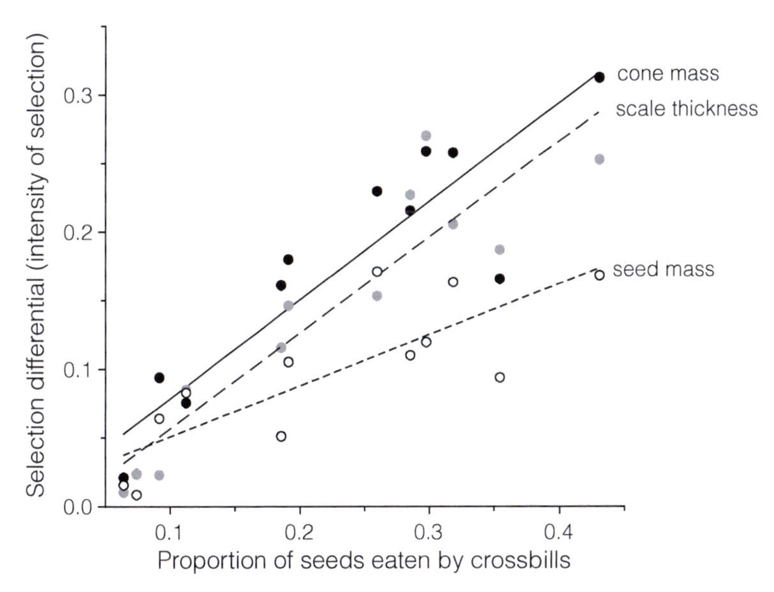

FIGURE 5.9 Selection differentials, which are standardized measures of the strength of natural selection, increase with increases in the proportion of seeds eaten by crossbills foraging on cones of five species of pines, including Rocky Mountain Lodgepole Pine. Each data point represents a different estimated selection differential, and different symbol types represent different traits: black circles = cone mass; gray circles = scale thickness; open circles = seed mass. The lines represent best-fit linear regressions. Modified from Benkman and Mezquida (2015).

it is doubtful that crossbills prefer to forage on cones with smaller Lodgepole Pine seeds per se. Rather, crossbills avoid larger cones with thicker distal scales and thereby avoid cones with larger seeds because seed size is positively correlated with scale size (McGinley et al. 1990; Benkman and Mezquida 2015). Presumably, the increasing rewards to crossbills from consuming larger seeds is countered by the increasing time and energy expended to access them. Thus, as scale size increased over pine generations, so did seed size as a genetically correlated trait. Notably, Summers et al. (2010) found similar cone preferences for crossbills foraging on closed cones of Scots Pine *Pinus sylvestris*.

We tested whether the deep bill of the Cassia Crossbill is an adaptation for foraging on the large, thick-distal-scaled cones by comparing their average bill depth to that predicted as optimal for foraging on cones from the South Hills (Fig. 4.8). Their average bill depth of 9.85 mm is similar to, but slightly smaller than, the predicted optimum of 10.00 mm (Benkman 2003). Though the predicted optimum matches the mean bill depth of 10.00 mm for the Cypress Hills crossbills, I would place more faith in the estimated size of the Cassia Crossbill, as it is based on a vastly larger sample (2,501 individuals versus 10). In retrospect, a smaller than 'optimum' bill depth is not surprising, because we used average-sized cones in the experiments to predict the optimum. Though a defensible choice at the time, we subsequently learned that Cassia Crossbills forage consistently on smaller than average-sized cones with relatively thin scales (Benkman et al. 2013). Consequently, a bill smaller than 10.00 mm deep is expected to be optimal for foraging on the cones normally utilized, although we don't know how much smaller.

The reciprocal adaptations between crossbills and Lodgepole Pine are visually striking (Fig. 5.2)—all the more so because they may have coevolved within only the last 5,000 years. Models of the geographic distribution of Lodgepole Pine based on estimated climatic conditions 6,000 years ago (Parchman et al. 2016) are consistent with paleobotanical studies (Mehringer 1985; Davis et al. 1986) and indicate that Lodgepole Pine was sparse in the South Hills/Albions. Thus, these ranges were unlikely to support a resident crossbill population during the dry mid-Holocene warm period from 5,000 to 7,000 years ago. In the Cypress Hills, the pollen fossil record shows that Lodgepole Pine became common only in the last 4,600 years, after the mid-Holocene warm period (Sauchyn and Sauchyn 1991; Sauchyn 1997). Lodgepole Pine apparently spread from the northern Rockies, arriving in the region of the Cypress Hills around 12,200–10,400 years ago (MacDonald and Cwynar 1985), but remaining sparse there until 4,600 years ago.

Although there are many similarities between the crossbills and Lodgepole Pine in the South Hills/Albions and Cypress Hills, the population dynamics of the crossbills most likely have differed. A crossbill population has probably persisted in the South Hills/Albions over the last 5,000 years because the main forest disturbance for Lodgepole Pine—stand-replacing fire—was unlikely to burn most or all the Lodgepole Pine within both ranges simultaneously. Prior to the introduction and spread of Cheatgrass *Bromus tectorum*, sagebrush steppe habitat—the ground cover surrounding most of the Lodgepole Pine patches within the South Hills/Albions and between them (Fig. 5.5)—was not a particularly good conduit for fire (Baker 2006). The genetic distinctiveness of the Cassia Crossbill further attests to the persistence of a crossbill population over time (Parchman et al. 2016).

Such crossbill persistence was unlikely in the Cypress Hills (even without the introduction of squirrels), which is surrounded by frequently burned prairie (Strauss 2001). Here, fires may have spread broadly enough to periodically burn all or nearly all the pine. For example, a fire in 1885 killed all but a few scattered pockets. A mere four years later the Cypress Hills burned extensively again (Strauss 2001). It thus seems likely that over the last 4,600 years, crossbill populations in the Cypress Hills were extirpated at least occasionally after extensive fire. As the forest and canopy seed bank developed after a fire, larger-billed individuals of formerly nomadic Red Crossbills may have become resident (performance-based habitat choice, as discussed in Chapter 3), and a locally adapted population could have originated quickly, perhaps with little subsequent evolution—although rapid evolution as found in Darwin's ground-finches (Grant and Grant 2014) could have also contributed. This could explain the population of large-billed crossbills found by Godfrey 60 years after the 1880s fires. As a result of such rapid local adaptation to the pine, the crossbills likely reached comparable densities and exerted similar levels of selection in each pine generation as in the South Hills/Albions, which would account for the parallel evolution in the cones. Future genetic analyses that include DNA from specimens collected in the Cypress Hills in the 1940s by Rand and Godfrey will help evaluate whether Cypress Hills crossbills were distinct genetically.

To summarize (Fig. 5.2): Red Squirrels are common throughout most of the range of Rocky Mountain Lodgepole Pine. They are superior competitors for seeds in the cones and important agents of selection against serotiny, reducing its frequency (Box 5.1). Because of a diminished canopy seed bank, crossbills are relatively uncommon in these forests. Selection exerted by Red Squirrels drives the evolution of cone structure, and Red Squirrels adapt to the enhanced defenses and coevolve with Lodgepole Pine. Crossbills adapt to the older closed cones that escape Red Squirrel harvest. But in some isolated mountain ranges east and west of the Rockies, Red Squirrels are absent, and were absent from the Cypress Hills until they were introduced in 1950. Here, pines lost their defenses directed at Red Squirrels. The absence of cone harvest by Red Squirrels and increase in occurrence of serotiny results in the massive accumulation of a canopy seed bank that supports high population densities of crossbills, especially in the largest ranges. Here, crossbills are the primary agent of selection on cone structure, favoring the evolution of larger cones with thicker distal scales. This in turn favors the evolution of crossbills with larger bills and a continuing escalation in the defenses of pines and the counter-offenses in crossbills, typifying a coevolutionary arms race that is replicated on both sides of the Rockies (Fig. 5.2).

Variation in the distribution of co-occurring species underlies the geographic mosaic of coevolution in many other studies, including variation in the co-occurrence of competitors (Craig et al. 2007; Siepielski and Benkman 2007), alternative hosts (Zangerl and Berenbaum 2003) and co-pollinators (Thompson and Fernandez 2006). Some studies have also detected mismatches between interacting species (Kaltz et al. 1999; Zangerl and Berenbaum 2003), as expected by theory (Thompson 2005). In crossbills, we have not. I suspect this is due in part to performance-based habitat choice (Chapter 3), a phenomenon where individuals settle where they are well matched to the cones and, at least in the case of the Cassia Crossbill, reproductive isolation prevents much gene flow (Chapter 7, 'Reproductive isolation of the Cassia Crossbill').

A coevolutionary arms race in Newfoundland

Finding evidence of coevolution in the South Hills caused us to reconsider the Newfoundland Crossbill (Fig. 5.1). I had assumed its evolution was simply a consequence of the loss of seed defenses directed at Red Squirrels (Benkman 1989c). Now that seemed unlikely. When I started thinking about revisiting Newfoundland, a promising new graduate student, Tom Parchman, arrived in my lab. He was keen on such a project; Tom helped with the study in the Pindos Mountains mentioned in Chapter 3, 'Cone structure', and has led our genetic studies on crossbills ever since. In September 2000, Tom traveled to Newfoundland and sampled Black Spruce cones from two islands along its northeast coast where Red Squirrels remained absent. Because Red Squirrels were so numerous on Newfoundland and harvest such a large fraction of the cones soon after they mature in early autumn (64–95%; West 1989), we felt that we needed to find areas without Red Squirrels to get a random sample of cones. We did not want to solely sample from trees that were avoided by that species. Tom also sampled cones at four sites in Quebec where Red Squirrels are native. Here on the mainland, we were not worried that their cone harvesting would bias our samples, as Red Squirrels harvest only a small fraction of the mainland cones (19% at most; Prévost et. al. 1988). Moreover, Tom was able to sample from sites where he did not detect evidence of Red Squirrels.

Much as we found when Red Squirrels are absent from Lodgepole Pine forests, Tom discovered that Black Spruce on Newfoundland has lost apparent defenses directed at Red Squirrels. His data suggested that the number of seeds per cone increased and the ratio of cone mass to seed mass decreased by 33% on Newfoundland relative to areas of Quebec where Red Squirrels are present (Parchman and Benkman 2002). However, Tom also found that potential defenses directed at crossbills are elevated on Newfoundland, just as we had found for Lodgepole Pine when Red Squirrels are absent. Scale thickness is 15% greater on Newfoundland than on the mainland; the comparable increase in distal scale thickness in Lodgepole Pine is 12%. Thus, Black Spruce had not simply lost squirrel defenses as I had proposed earlier (Benkman 1989c). Black Spruce had also evolved increased defenses directed at crossbills, indicative of a coevolutionary arms race between the two species.

In addition, Tom measured the feeding intake rates of captive Red Crossbills foraging on Black Spruce cones from all the sites from which he had collected cones. His results supported the interpretation that Black Spruce in Newfoundland has evolved elevated defenses against crossbills favoring larger bill sizes than on the mainland. However, Tom could not conclude that the Newfoundland Crossbill matched the optimum bill depth, for the simple reason that the range of bill depths of his captive crossbills did not encompass that of the massive Newfoundland Crossbill. Nonetheless, all his results were consistent with the hypothesis that Newfoundland Crossbills and Black Spruce had been engaged in a coevolutionary arms race and thereby provide another example of the geographic mosaic of coevolution (Parchman and Benkman 2002).

Coevolutionary interactions elsewhere

The consistency with which crossbills, Lodgepole Pine and Black Spruce evolved in apparent response to the presence and absence of Red Squirrels (Figs 5.1 and 5.2) indicated that evolution was repeatable. If there are no Red Squirrels, then crossbills increase in number and coevolve with the conifers. But was this also the case in regions where *other* species of tree squirrels coexisted with *other* pine or spruce species? As discussed in Chapter 2, tree squirrels in the genus *Tamiasciurus* are confined nearly exclusively to North America north of Mexico, whereas tree squirrels in the genus *Sciurus* are widespread in North America south through Mexico and Central America, as well as in Eurasia. I was therefore enthused when Eduardo Mezquida contacted me about evaluating whether Red Crossbills coevolve with Aleppo Pine *Pinus halepensis* in the absence of tree squirrels in his native country of Spain (Fig. 5.10). Eduardo had recently completed his PhD in Argentina and proposed to compare Aleppo Pine cones from the Iberian Peninsula, where European Red Squirrels *Sciurus vulgaris* are widespread, to cones on two of the Balearic Islands where the squirrels are absent. The fact that a subspecies of Red Crossbill *Loxia c. balearica* is endemic to Mallorca (Fig. 5.10), the largest of

FIGURE 5.10 Aleppo Pine *Pinus halepensis* and Red Crossbill *Loxia c. balearica* on Mallorca. (A) A cone that had been foraged on by a crossbill. (B) A female captured on 22 September 2007. Aleppo Pine (C) is widespread in mountainous areas of Mallorca including on Cap de Formentor (D).

the Balearic Islands, suggested something different was occurring there. Eduardo secured funding and joined us in Las Cruces, then he returned to Spain to conduct fieldwork.

Sciurus harvest and cache fewer cones than do *Tamiasciurus*, so I did not expect species in the genus to have as strong a competitive effect on crossbills or as strong a selective effect on cone structure. Nevertheless, I expected there to be differences in cone structure between areas with and without *Sciurus*, similar to what we had found in such comparisons for *Tamiasciurus*. To my surprise, Eduardo didn't find the same patterns. *Sciurus* preferentially foraged on smaller, thinner-scaled Aleppo Pine cones and favored the evolution of larger cones with thicker scales (Mezquida and Benkman 2005). Because this is similar to the selection exerted by crossbills (Fig. 5.9), cones in areas with both *Sciurus* and crossbills (Iberian Peninsula) have more features deterring crossbills than areas where *Sciurus* are absent (Balearic Islands). The cones on Mallorca were smaller with thinner (and shorter) scales, which presumably account for why the endemic crossbill there has a small and relatively short bill (Fig. 5.10B) compared to crossbills on the Iberian Peninsula associated with Aleppo Pine (Alonso et al. 2020). However, this comparison alone does not provide a test of whether the Aleppo Pine evolves in response to selection exerted by crossbills, or whether both species are coevolving on Mallorca. The smaller cones with thinner scales in Mallorca could simply be the result of relaxation of selection in the absence of *Sciurus*, with no effect from crossbills.

Fortunately, Eduardo gathered cones from Ibiza, another Balearic Island where *Sciurus* are absent. This provided a comparison between two nearby islands without *Sciurus*, only one of which—Mallorca—has a long history with crossbills. The only notable difference between the two islands was the presence of thicker scales on cones on Mallorca, as expected from selection exerted by crossbills (Mezquida and Benkman 2005). So, though the endemic Mallorca crossbill has a small bill compared to crossbills on the Iberian Peninsula, its bill would presumably be smaller still if selection by crossbills on Mallorca had not favored thicker cone scales. Thus, we once again found evidence that crossbills coevolve with pine in the absence of tree squirrels. Subsequent studies revealed evidence of coevolution in the absence of tree squirrels between Hispaniolan Crossbills *Loxia megaplaga* and *Pinus occidentalis* in Hispaniola (Parchman et al. 2007), and an endemic Red Crossbill subspecies *Loxia c. guillemardi* and European Black Pine *Pinus nigra* on Cyprus (Benkman and Parchman 2009).

One of Eduardo's intriguing results is that tree squirrels preferentially harvest longer cones from trees when their cones average under 7 cm long, but preferentially harvest smaller cones from trees when their cones average over 7 cm long. This suggests that tree squirrels preferentially harvest Aleppo Pine cones of an intermediate length of around 7 cm. Although longer cones have more seeds and hence provide greater total energy rewards, it is likely that cone handling time increases exponentially with increasing size, as found for seed handling times in relation to seed size for seed-eating birds (Benkman and Pulliam 1988). The hypothesized relationships are illustrated in Figure 5.11. Although the exact relationships and optimal cone sizes for tree squirrels will likely vary depending on cone structure and squirrel size, we should expect that tree squirrels preferentially harvest the larger cones of conifers that produce small cones and preferentially harvest the smaller cones of conifers that produce large cones. This is indeed what we find: for example, American Red Squirrels preferentially harvest the largest cones of Lodgepole Pine, whose cones average less than 4 cm long; comparisons of Black Spruce cones from Newfoundland to those in Quebec suggest a preference for larger in the even smaller cones of Black Spruce. We find similar results for two species of *Sciurus* (Abert's Squirrel *S. aberti* and the Western Gray Squirrel *S. griseus*) harvesting Ponderosa Pine as Eduardo did for Aleppo Pine, with the optimal cone length for squirrels approximating 8 cm (Parchman and Benkman 2008). Likewise, size preferences in Western Gray Squirrels and perhaps Douglas Squirrels *Tamiasciurus douglasii* presumably account for why the already relatively large-coned pines of the west coast of North America tend to have even larger cones where the pines occur with tree squirrels (e.g., Linhart 1978).

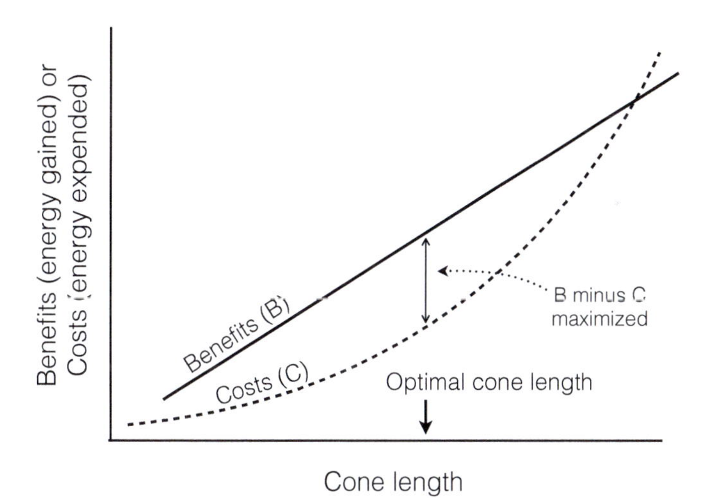

FIGURE 5.11 The energy reward to tree squirrels foraging for seeds in conifer cones increases roughly linearly with increasing cone length, because of a concomitant increase in the number of seeds. The handling cost to tree squirrels will also increase with increases in cone length, but most likely in an accelerating manner as found for increases in seed size for seed-eating birds (Benkman and Pulliam 1988). At shorter cone lengths, benefits increase faster than costs, such that longer cones should be preferred. At longer cone lengths, costs increase faster than benefits, so that shorter cones should be preferred. The optimal cone length is where the benefits most exceed the costs (B minus C is at a maximum), which occurs where the slopes of the two curves are equal. The result is that natural selection exerted by tree squirrels will tend to cause small-coned conifers to evolve even smaller cones and large-coned conifers to evolve even larger cones.

Birds versus mammals

A reoccurring theme of our research beginning with my thesis on nutcrackers, Red Squirrels and Limber Pine (Benkman et al. 1984) is the apparent dominating effect of tree squirrels on the ecology and evolution of conifer forests. This impact became even more fascinating one evening in 1994 while having dinner with the eminent mammologist and ecologist Jim Brown. Earlier in the day I had given a seminar on my research at the University of New Mexico, where Jim is now an emeritus professor. Jim mentioned that for any given body size, the population densities of mammals averaged about an order of magnitude greater than that for comparably sized birds—something I had been completely unaware of. I later saw this pattern illustrated in his textbook *Macroecology* (Brown 1995) and in a subsequent paper (Silva et al. 1997), both of which plotted population densities of different species of birds and mammals in relationship to their body size. These differences are remarkable, as birds and mammals are both endothermic with roughly similar mass-specific metabolic rates, such that an individual of a similar body size uses similar amounts of energy per day. A further indication of an apparent ecological advantage of mammals over birds is the approximate twofold higher probability of establishment by mammals than birds when introduced into new environments (Jeschke 2008).

Why is there such a large difference in the population density and invasive ability between the two groups? One factor must be the much-reduced small intestine and its surface area in birds relative to non-volant mammals of similar body size and diet (Lavin et al. 2008). This difference reduces both the absorption of nutrients and the competitive ability of birds. Regardless of its cause, the ecological impact in terms of energy demand and flux in a given community could be nearly an order of magnitude greater for an average species of mammal than for a comparably sized species of bird. The implication is that mammals will generally have a greater ecological and likely greater evolutionary impact than birds, much as I have found for tree squirrels relative to crossbills. Indeed, a number of studies are consistent with the broader hypothesis that mammals have especially great ecological impacts (Kalka et al. 2008; Williams-Guillén et al. 2008; Bello et al. 2015; Sobral et al. 2017). Furthermore, once introduced and established, the frequencies at which impacts on other species from herbivory and predation are detected are over an order of magnitude greater for introduced mammals than for introduced birds (Ebenhard 1988). There are few examples from which to assess the evolutionary impact of mammals relative to birds. The uniqueness of New Zealand, from which non-volant mammals were absent for many thousands of years, hints at their potentially massive influence on the evolution of local biotas. In particular, the long absence of non-volant mammals from New Zealand is why Tim Flannery (1994: 55) described New Zealand as 'a completely different experiment in evolution from the rest of the world'.

I am thus haunted by Robert Whittaker's comment that 'Birds seem to be evolution's ornaments on the tree of community function' (Whittaker 1977: 42). Of course, I don't mean to discount the significance of birds. They are clearly important for numerous ecological and economic reasons (Sekercioglu et al. 2016). But perhaps most of all, birds have been invaluable subjects in advancing our understanding of biology (Konishi et al. 1989)—along with being the sparkle that for many of us brings 'the tree' to life!

Chapter 6

Flocking, Patch Assessment and the Evolution of Contact Calls

> Particular calls are, I believe, important signals in securing reproductive separation between the two forms [Red and Scottish Crossbills].
>
> (D. Nethersole-Thompson 1975: 25)

Having found evidence that the common call types of Red Crossbills *Loxia curvirostra* in western North America are each adapted for foraging on a different conifer species (Chapter 4), I wondered what keeps the different ecotypes distinct and to what extent they are reproductively isolated. Might they represent species as Groth (1993a) advocated, or something else? Most ecotypes are not morphologically discrete—they overlap extensively in size with other ecotypes—so that many individuals cannot be identified to ecotype by simply measuring them. Neither do the ecotypes differ consistently in plumage (Groth 1993a). Most individuals need to call for us to categorize them! What keeps an individual from one ecotype whose morphology is closer to the average of another ecotype than to its own from foraging and associating with the other ecotype, especially when they wander widely and commonly co-occur? In fact, you might expect that it would benefit an aberrant individual to associate with a different ecotype than its own, when its bill structure is closer to the optimum for the cones of a different species—a scenario that could result in mating and a breakdown of barriers between lineages that might otherwise go on to become species. In the face of such possibilities, what might limit gene flow between ecotypes that commonly breed in the same forest, and allow them to evolve independently as they adapt to foraging on different conifers?

Just as we use their contact calls to identify them, it is conceivable that crossbills do the same—that these calls provide the social identity (or 'glue') that keep ecotypes distinct, a hypothesis put forward by Nethersole-Thompson (1975) in the epigraph above. Given the similarity in plumage and only subtle differences in average size among various co-occurring ecotypes, visual cues might be of limited use. This is especially true when decisions to join other individuals are made at a distance. In contrast, contact calls, which serve to coordinate flock movements, could act to distinguish ecotypes, much as we use their calls to identify them.

Contact calls are learned from parents so that, like bill size, they are inherited (Groth 1993a; Keenan and Benkman 2008; Sewall 2011) and potentially provide a marker of average bill structure (Snowberg and Benkman 2007). Numerous observations of foraging flocks of crossbills (Fig. 6.1) suggested to me why such distinct contact calls might be favored as badges or markers for identification. These observations also led to ideas on how flocking behavior could contribute to assortative mating and reproductive isolation, which is the focus of the next two chapters on speciation. Here I focus on flocking behavior and how accruing certain benefits from flocking requires that flocking occurs among individuals with similar bill sizes (i.e., is assortative).

Public information and assortative flocking

A flock of foraging crossbills generally makes few sounds besides the cracking of cone scales and the occasional quick flutter of wings while flying between branches or trees. This can be enough to draw your attention. However, sometimes foraging flocks are so quiet that I've felt that I detected them with a sixth sense, although surely I was alerted by a slight movement or sound. The quiet ends though when one or two crossbills call. Sometimes the calling stops, and the crossbills continue foraging. Sometimes others join in with calling, reaching a crescendo that ends

FIGURE 6.1 White-winged Crossbill *Loxia l. leucoptera* flock composed of two reddish males and a yellowish female foraging on reclosed White Spruce *Picea glauca* cones. Crossbills occur in flocks year-round, especially when foraging and including while nesting. Photograph taken by Justin Della Mora Duquette in Montreal, Quebec, Canada on 20 February 2022.

with the flock flying off. This preamble before departure has been noted by others including William Brewster, who wrote: '[T]hey are absolutely silent when feeding. Just before starting to fly one or two birds begin to call, others join in and finally with a general outcry the flock are [is] off' (quoted in Taber 1968: 537). The increase in calling or a high intensity of calling might signal some level of consensus among flock members to depart.

Increased calling as in crossbills as well as distinctive vocalizations have been found to coordinate the timing of group departure in a diversity of birds and mammals (Sperber et al. 2017); although this requires quantification for crossbills with additional data using methods such as those applied by Dibnah and colleagues (2022) for Jackdaws *Corvus monedula*. Such coordination facilitates group cohesion, thereby enabling individuals to reap the benefits of grouping more often. These benefits include reduced risks of predation from the dilution effect (i.e., the probability that any one individual is attacked and killed decreases with group size), enhanced predator detection from increased collective vigilance, and less time scanning for predators and thus more time available to forage—all of which have been found in numerous groups of animals (Krause and Ruxton 2002). In White-winged Crossbills *Loxia l. leucoptera*, I've detected evidence for all but the dilution effect, which is undoubtedly still occurring, though more difficult to prove (Benkman 1997).

I first started thinking that calls served other purposes than simply coordinating flock departure while measuring intake rates in the field. One Eastern Hemlock *Tsuga canadensis* stands out in my memory. It was 26 February 1985 in Algonquin Provincial Park, Ontario. This hemlock towered above the surrounding trees. Large numbers of White-winged Crossbills arrived in the summer of 1984, part of a late spring to early summer movement from Alaska and the Yukon described in Chapter 2, 'Crossbill responses'. In late February 1985, white-wings seemed to be moving again, with passing flocks often drawn to the tall hemlock. In spite of this, it had few seeds in its cones, resulting in low seed intake rates for the crossbills. Within perhaps 10 or 20 seconds after starting to forage a single individual would begin calling. Others soon joined, with the calling rapidly reaching a crescendo, before off they flew. It was as if the first to call was saying 'I'm doing poorly, how are you doing?' Others in turn would answer by calling, implying that they too were doing poorly. And if most were doing poorly, then each flock member could be assured that unprofitable cones were not limited to the few each crossbill had sampled. Rather, they were representative of cones throughout the tree. In other trees, one crossbill might begin calling, but the rest of the flock remained silent and continued foraging. The caller then quieted and continued foraging. The caller may have foraged on one or several cones with few or difficult to access seeds, but the silence of the rest of the flock implied that such cones were atypical for the tree.

I thought of testing whether a crescendo of calls caused a flock to depart even from a tree with lots of seeds in its cones by setting a speaker in a tree canopy and playing a recording of a crescendo of calls. In the summer of 1987,

I attempted to follow through on this experiment, nailing wooden planks up Tamarack *Larix laricina* trees in an area where white-wings were common, so that I could access multiple canopies to measure the number of seeds in the cones and to place a speaker to conduct playbacks. The project didn't get much further, however, because upon counting the number of full seeds in the cones, I realized that seeds were so few that the crossbills were unlikely to remain for long—and they didn't.

As the years went by, I thought more about the use of foraging information from flockmates. I had become increasingly interested in the diversification of crossbills (Chapters 4 and 5), and I began to envision a mechanism by which flock foraging could contribute to reproductive isolation as Red Crossbills adapted to alternative conifers. I reasoned that if crossbills use the intake rates of flockmates to help assess patch quality (Clark and Mangel 1984, 1986; Valone 1989, 1993), then it would be advantageous to flock with individuals having similar bill sizes and hence similar foraging abilities. Conversely, if crossbills flocked randomly in terms of bill size, then the feeding performance of flockmates should be less useful and might even be misleading. A cone type that is profitable to a large-billed crossbill is often unprofitable to a small-billed crossbill and vice versa. In contrast, a cone type that is profitable to a large-billed crossbill will be similarly profitable to other large-billed crossbills. This would favor assortative flocking by bill size if flockmates are to provide useful information on tree quality. Assortative flocking would also reduce conflicts of interest among flock members about patch departure decisions (Conradt and Roper 2005), so that individuals remained together for longer with more opportunities to accrue the various benefits from flocking (Hutto 1988; Valone 1993). Assortative flocking in turn has consequences for reproductive isolation, because crossbills flock year-round and appear to choose mates from within flocks (Newton 1967, 1972; Nethersole-Thompson 1975). In particular, assortative flocking should lead to assortative mating and reproductive isolation between groups of crossbills differing in bill size. This assumed bias toward mating with flockmates has been documented in another cardueline finch, the Eurasian Siskin *Spinus spinus* (see Fig. 1.2), where in mate choice experiments, females preferred male flockmates even over males with brighter plumage that were not flockmates (Senar et al. 2013).

The challenge was to first test whether crossbills used the observed or communicated success of flockmates to make their own foraging decisions. If so, a natural follow-up hypothesis was that this information would be helpful when flockmates had similar feeding abilities but not when they had dissimilar feeding abilities. I raised this idea with Julie Smith, a graduate student who joined my lab in 1995. However, it was not immediately obvious how best to design the experiment. I knew we had to use captive crossbills, but using crossbills of sufficiently different bill sizes to get consistent differences in intake rates would be a challenge. Fortunately, the aviaries I had built at New Mexico State University inspired a solution.

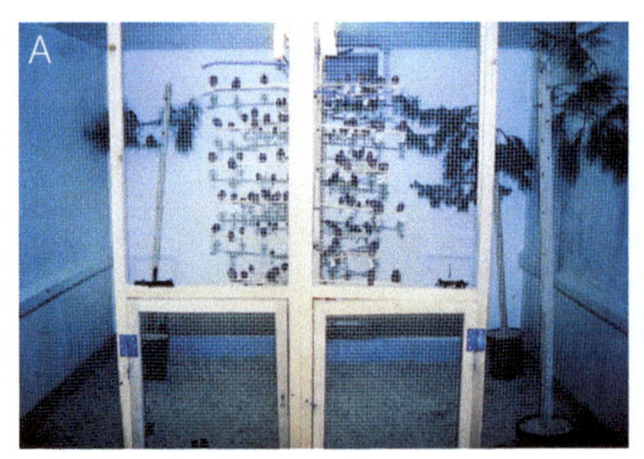

FIGURE 6.2 Photographs of the aviary setup where each half of two artificial trees was attached to opposite sides of the screen (hardware cloth) partition. At the beginning of a trial, crossbills were released into the aviary through small screen doors in the front of the aviary (A). The black strips of Velcro used to close the doors are near the top of the horizontal beams of wood. The branches with needles on the sides of the aviary provided perches, but no seeds. Four Lodgepole Pine *Pinus contorta* cones were attached and arranged in a square layout near the end of each branch (B) for a total of 40 cones on each tree half. Crossbills always landed first on the tree nearest the front of the aviary (A). In B, the focal bird is on the near side and its flockmates are on the other side of the partition.

The aviaries we used for holding crossbills and conducting foraging experiments were in rooms divided down the middle by hardware cloth—a wire mesh forming half-inch (1.3 cm) squares (Fig. 6.2). We used the hardware cloth dividing the room to partition in half two artificial trees made of dowels (Fig. 6.2B). Each half of a tree had 10 wooden dowels ('branches') with shorter dowels attached near their tips forming a cross. Three Lodgepole Pine *Pinus contorta* cones were attached near the tips of the dowels on each branch and a fourth was attached along the main dowel, forming a square with a total of 40 cones on each tree half.

All the cones were open and seeds were placed at the base of the scales, requiring crossbills to perch next to or on the cone so that they could peer between the scales to locate seeds. Once located, seeds were removed quickly regardless of bill size, thus we could easily standardize intake rates by standardizing the number of seeds added to the cones. This design also allowed us to set up quickly so that we could run multiple experiments in a morning.

Because flockmates could see but not interfere with each other through the screen (Fig. 6.2B), we eliminated physical displacement and competitive scramble. In response to the latter, birds might forage faster because flockmates deplete the patch (Clark and Mangel 1986; Shaw et al. 1995). Importantly, we could control the number of seeds on each side of a tree independently and thereby control the average intake rates of crossbills on each side. Thus, we could determine whether foraging with flockmates that have similar intake rates enhanced foraging performance relative to when foraging alone. We also could test whether dissimilar intake rates among flockmates depressed foraging performance relative to when foraging in flocks comprised of individuals with similar intake rates. In an earlier study in

New Brunswick, Canada (Benkman 1997), I found that White-winged Crossbills spent less time looking for predators when foraging with a flockmate than when alone, and thus could remove and eat seeds more quickly from cones. To control for such an effect when comparing solitary foragers to those in flocks, a crossbill was placed on the other side of the partition without foraging opportunities during the solitary trials. This crossbill provided another set of eyes for the solitary forager but provided neither foraging information nor interference.

To begin, we initially examined whether solitary foraging crossbills in our experiment approximated the optimal strategy for leaving a tree. If their foraging behavior approximated the optimum, then this would indicate that the setup was not too contrived and represented a reasonable challenge in which crossbills were attempting to maximize their energy consumption. As in foraging trials previously discussed, the crossbills had gone without food since the previous afternoon, so as to motivate them to forage quickly and consistently. In the experiments, one tree (the tree half on the focal bird's side) had three seeds in each of four randomly selected cones, with the remaining 36 cones empty. The other tree had no seeds, and which tree had seeds was determined randomly. Although we suspected that crossbills could distinguish more subtle differences in intake rates than seeds versus no seeds, our goal was to test whether crossbills used the occurrence or presence and absence of seed husking by flockmates to help assess when to leave one tree and search in the other. This allowed for a more manageable and repeatable design, while nevertheless testing for whether crossbills used the feeding success of flockmates to assess food patches.

Crossbills always landed first on the tree nearest to the release points, which was empty half the time. Crossbills searched the tree to determine if it had seeds and then flew to the back tree to search it. Julie and her team (that sometimes included me) terminated the trial when the focal crossbill stopped searching on the second or 'far' tree. When the near tree was empty, the question of interest was whether crossbills searched the optimal number of cones before departing. If too few cones were searched, crossbills might leave a tree with seeds without finding any of them. And if crossbills visited too many cones, crossbills were wasting their time. In the latter case, they had searched more than enough cones to be confident the tree was empty. This is analogous to the problem discussed in Chapter 1, 'When to leave a cone for another?'.

After a series of training trials for each of five Red Crossbills of the Ponderosa Pine ecotype, the number of cones a crossbill visited before leaving an empty tree stabilized, presumably because the crossbills learned the distribution of seeds and settled on a cone sampling routine. Using data Julie gathered, we solved for the optimum number of cones to visit on the empty tree before leaving it using a mathematical model for patch sampling behavior developed by Steven Lima (Lima 1985). For four of the five crossbills the number of cones visited was reasonably close to each bird's predicted optimum of visiting between 16 and 20 cones before departing the empty tree; sampling was within 90% of the optimum in terms of maximizing intake rate. The fifth crossbill visited on average 32 cones when its predicted optimum was to visit 20 cones. We later tested this crossbill after it had gained additional experience and it visited on average 18 cones, well within 90% of its optimum. Thus, we felt it was safe to assume that the setup was not too contrived for the crossbills. Moreover, we had a baseline for judging whether having flockmates might enhance their patch or tree assessment.

We then paired the crossbills with one flockmate that foraged on the other side of the partition to determine whether crossbills landing on a tree without seeds would visit fewer cones before leaving than when foraging alone. Somewhat surprisingly to us, crossbills foraging with a foraging flockmate in our experimental setup (one tree empty and one tree with seeds) did not seem to alter their behavior and visited a similar number of cones before leaving an empty tree as they did when foraging alone. That is, crossbills didn't seem to pay attention to the foraging success of a single flockmate.

To test further whether crossbills use the feeding performance of flockmates, we increased the number of flockmates to two to provide a greater opportunity for the focal crossbill to observe foraging success or failure of flockmates. We found that the focal crossbill left empty trees after visiting only about half as many cones than when foraging alone. There was also a trend for focal crossbills to eat more of the seeds on full trees before leaving them. So, there was no cost of searching fewer cones on empty trees, only a benefit. Because the focal bird was the first bird to leave the empty tree over 20% of the time, it indicated that it wasn't simply following the other birds, although such departures provide information on tree quality too if flockmates assess the patch similarly.

The results for flocks of three but not two (Box 6.1) indicate that crossbills used, in addition to their own personal foraging information, the foraging behavior of flockmates to assess food patches more rapidly. The obvious cue or so-called public information (Valone 1989) that crossbills likely used is the distinctive behavior of crossbills husking seeds, which involves lifting their head from the cone to mandibulate seeds for several seconds to remove and discard the seed wings and seed coats and then swallow the kernels (Fig. 1.1). We thought of placing an opaque sheet between the two halves of the trees to remove such visual cues to perhaps elicit calling as described at the beginning of this chapter. However, we never did this, in part because the scale of the setup—both the size of the trees and the time taken to search them as well as the short distance between them—were unlikely to necessitate or favor calling. Calling nevertheless has advantages over visual cues: both silence (implying that the patch is worth sampling further) and calling (indicating it is not) can be more readily acquired from a greater number of individuals, especially in an obstructed environment like the canopy of a conifer. Furthermore, crossbills can then focus on foraging and scanning for predators rather than peering at flockmates.

Upon finding evidence that flockmates can enhance an individual's assessment of resource patch quality, we set out to see if such benefits vanish when flockmates

Box 6.1: Benefits of a quorum

We were surprised that crossbills didn't seem to use foraging information from a *single* flockmate. However, because the accuracy of information provided by two individuals is greater than that provided by one, it is conceivable that an individual might use public information only when it can acquire information from a quorum of individuals (Sumpter and Pratt 2009). Indeed, evidence for the requirement of a quorum to instigate patch departure has been found in Meerkats *Suricata suricatta*, where they are much more likely to move to a new food patch when three individuals give their 'moving' calls than when only two individuals give such calls (Bousquet et al. 2011). Bousquet and colleagues also note other examples where signals from two to three individuals are needed to instigate group movements.

Nevertheless, in our experimental setup, the foraging success of a single flockmate should have proved useful. I suspect that our finding that crossbills seemingly ignored the foraging success of a single flockmate reflects their evolution in a less contrasting environment (i.e., not as extreme as seeds versus no seeds) where the information provided by the intake rates of a single flockmate might not be that reliable. In nature, intake rates of individuals even with identical abilities are likely to vary among cones within a tree, and average differences between trees are generally more subtle than in our experimental setup. This may have favored crossbills that discount public information on feeding performance when they have only one flockmate. It is also in this more subtle environment that a quorum of more than two individuals might be especially valuable for improving not only the speed of patch assessment, as we found, but also its accuracy (Sumpter and Pratt 2009). It would be interesting to determine whether the speed and accuracy of patch assessment by crossbills increases with increases in flock size beyond three individuals, especially in an environment with more subtle differences among patches.

have dissimilar intake rates. We used the same experimental design as before except for the fact that now, if the near tree for the focal crossbill was empty, then the *other* side of the tree had seeds (and vice versa). Thus, when the focal crossbill landed on the near tree without seeds its flockmates would find seeds, and conversely when its half of the near tree had seeds its flockmates would not find seeds. This provided a strong contrast between the intake rates of flockmates on a given tree, but would be within the realm of differences between those of two ecotypes foraging on the closed cones of some conifers.

The results were striking. When the near tree was empty for the focal bird, it often seemed to notice its flockmates husking seeds on the other side of the tree. In apparent response, the focal bird continued searching cones and thereby visited many more cones than would be adaptive. Instead of visiting only 10 cones and rarely revisiting cones when both sides of the tree lacked seeds, the focal bird remained much longer, on average visiting between 20 and 60 cones and oftentimes revisiting them. Conversely, when the near tree had seeds on the focal bird's side but none on the other side, the focal bird visited fewer cones and ate fewer seeds prior to leaving the tree (usually about 4 seeds but sometimes 0 seeds eaten, versus 9 to 10 seeds eaten when both sides of the tree contained seeds). This left little doubt that crossbills use 'public' information, and in particular the feeding success of flockmates, to aid in assessing trees. Furthermore, the benefits of such information accrue only when flockmates have similar feeding abilities.

Does the propensity to flock together decrease as differences in bill size increase?

If crossbills benefit from having flockmates whose feeding abilities are similar to their own, then crossbills should be less likely to associate with individuals of another ecotype (so-called heterotypics) than of their own ecotype (homotypics). This seemed to be true from my observations in the field, although I hadn't quantified them. However, this isn't a particularly unique prediction: preferences for homotypics over heterotypics are found more generally. The more subtle and revealing prediction is that the extent to which crossbills discriminate against heterotypics should increase as the differences in bill depth between ecotypes increases. Individuals with greater differences in bill depth are expected to have greater differences in their intake rates on most conifers and provide less useful and more misleading foraging information.

We estimated the quality of public information provided by heterotypic flockmates as compared to that provided by homotypic flockmates using data we had at hand. We used the distribution of bill depths of 2,501 adult Cassia Crossbills *Loxia sinesciuris* to simulate differences in bill depth between individuals both within and between ecotypes, and we used the intake rates of crossbills foraging on Lodgepole Pine (Fig. 4.4A) to estimate how differences in bill depth affect absolute differences in intake rates between individuals (Porter and Benkman 2019). Figure 6.3A shows how our estimates of the relative quality of foraging information (i.e., similarity of expected intake rates) provided by heterotypics declines as compared to that provided by homotypics as average differences in bill depth between ecotypes increase. Of course, the relationship depends on how intake rates vary with bill depth. If intake rates do not vary with differences in bill depth, then heterotypics are likely to provide relatively high-quality foraging information. I address such situations in Chapter 8. However, we suspect that assortative flocking is favored when feeding on key conifers and increasingly so as the difference in bill depth increases between ecotypes. The question therefore

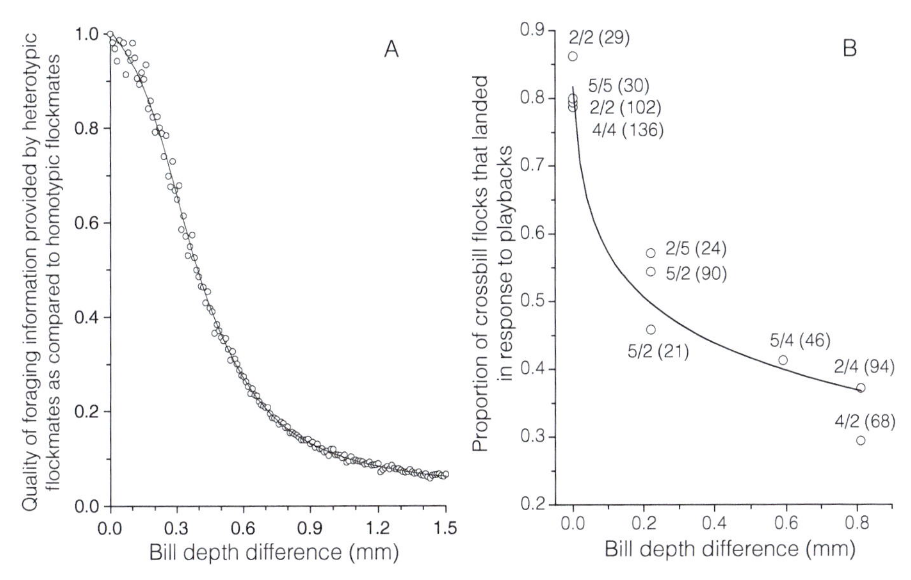

FIGURE 6.3 (A) The predicted quality of foraging information provided by heterotypic flockmates as compared to that provided by homotypic flockmates declines as the average difference in bill depth between ecotypes increases. From Porter and Benkman (2019). (B) About 80% of the free-flying flocks landed in response to playbacks of contact calls from their own ecotype, but the percentage landing decreased when the playbacks were of heterotypics and as the difference in average bill depth increased. Numbers to the left of diagonal lines above the data points (open circles) represent the ecotype of the playback, numbers to the right of diagonal lines represent the ecotype of the free-flying crossbills and numbers within parentheses represent the number of free-flying flocks. Data are from Smith et al. (2012) and Porter and Benkman (2022).

is whether the propensity to flock with heterotypics decreases as the difference in bill depth increases between ecotypes, as we predict (Fig. 6.3A).

Julie Smith and I had discussed testing for assortative flocking while she was a graduate student, and she even conducted a field study using live decoys of two ecotypes in 1998. However, it wasn't until over 10 years later and after Julie had moved to a faculty position at Pacific Lutheran University in Washington that Julie conducted more definitive experiments using playbacks of contact calls (Smith et al. 2012). I initially doubted that crossbills would be sufficiently responsive to playbacks, but Julie proved me wrong. She conducted the experiments on the east side of the Washington Cascades where Ponderosa Pine *Pinus ponderosa* and Douglas-fir *Pseudotsuga menziesii* were producing large seed crops, and Ponderosa Pine and Douglas-fir ecotypes were common and nesting. Julie also had the aid of two undergraduate students, Stephanie Sjoberg and Matthew Mueller. Here were two ecotypes foraging on two contrasting conifers where the relative quality of foraging information from heterotypics was expected to be poor—indeed, very poor, because the average difference in bill depth between these two ecotypes is 0.8 mm (Fig. 6.3A). Julie and her team used recordings of the two common ecotypes plus those from the Lodgepole Pine ecotype for playbacks.

The team recorded ecotype identity of free-flying crossbills (all 663 flocks that they recorded were comprised of just one ecotype, indicating strong assortment), as well as whether they continued flying or landed in trees near the playback speakers in response to standardized recordings of the different ecotypes. The results Julie and her team found were striking (Fig. 6.3B; this includes additional playback data from Porter and Benkman 2022). The decline in the propensity to land as the difference in bill depth increased (Fig. 6.3B) was notably similar in form to the expected decline in information quality (Fig. 6.3A). These results, along with those provided in Chapter 8, 'Assortative flocking and behavioral isolation between ecotypes', provide strong support for the idea that foraging performance of flockmates is important to crossbills in guiding flocking decisions, and that crossbills increasingly avoid flocking with heterotypics as their expected intake rates increasingly differ from their own. The alternative hypothesis that the propensity of flocking decreases with decreasing call similarity can be eliminated, because call similarity is unrelated to differences in bill depth (Smith et al. 2012).

How might new ecotypes evolve?

Variation in call structure among individuals within an ecotype (Groth 1993a) is the raw material for call evolution.[3] Much of this variation might arise simply because crossbills imitate the calls of their parents imperfectly (non-genetic inheritance; Groth 1993a; Keenan and Benkman 2008; Sewall 2011). Imperfect imitation increases variation, much like mutations increase genetic variation. Crossbills also modify the fine structure of their contact calls over their lifetime (Porter and Benkman 2019), although most call modification within adults might be related to changes occurring when crossbills imitate the call of their mate (Keenan and Benkman 2008; Sewall 2009). If call matching involves mostly call convergence between mates (Sewall 2009), this would tend to decrease population-level variation in call structure. Thus, imperfect call imitation of parents by offspring is likely to increase call variation, while call matching by mates potentially acts to reduce variation. On balance, these two processes should influence the overall level of variation. Nevertheless, some call variants are rather different and difficult to categorize into known ecotypes (Groth 1993a). Other variants might represent an accentuation or an addition or subtraction of an element within the original call.

Such variation could lead to the evolution of new ecotypes if these variants arise for example in geographic isolation, such as after crossbills colonize an island with a conifer that causes divergence in bill traits. However, new ecotypes need not arise in geographic isolation once a variant call arises if by chance (e.g., cultural

3 Although not all call types necessarily represent ecotypes, I refer to them here as such.

drift) it becomes associated with crossbills at one end of the bill size spectrum. For example, a variant call associated with only large-billed crossbills could be favored by providing these individuals a feeding advantage (Smith et al. 2012). Large-billed crossbills giving the variant call would have a feeding advantage because, given the propensity of crossbills to associate with individuals producing similar calls (Fig. 6.3B; Snowberg and Benkman 2007), they would generally flock with other large-billed crossbills. In contrast, large-billed crossbills producing the original call also given by small-billed birds would be at a relative feeding disadvantage, because they might flock just as often with small-billed as with large-billed crossbills. This relative advantage to individuals with the new variant call would be especially great when morphological differences are large, such as when the population relies on two (or more) species of conifers and experiences divergent selection leading to an increase in bill size variation. As Smith et al. (2012: 4227) noted: 'The increased feeding efficiencies of large-billed individuals with the variant call should translate into reproductive advantages and an increase in the number of large-billed offspring with variant calls because bill size is highly heritable in crossbills ... and crossbills imitate the contact calls of their parents.' Individuals with variant calls will eventually dominate the subpopulation of individuals with large bills. Then the original call will be associated with mostly smaller-billed crossbills, thereby increasing their foraging efficiency and fitness.

The advantages of assortative flocking should favor further differentiation between the original and variant calls if it allows crossbills to better discriminate among individuals, thereby increasing assortative flocking with like morphologies and within nascent ecotypes. Indeed, evidence for ongoing call divergence has been found in Cassia Crossbills (Porter and Benkman 2019). We first became aware of such a possibility when Patrick Keenan analyzed the contact calls of 114 Cassia Crossbills recorded multiple times over a maximum of five years (Keenan and Benkman 2008). Patrick had done nearly all the netting and banding of Cassia Crossbills over the previous three years, so he had recorded most of the birds he analyzed. He found that adult Cassia Crossbills subtly modify their calls over time in a manner that reduces their similarity in structure to those of the most similar-sounding, co-occurring ecotype (Ponderosa Pine ecotype). This trend, although consistent, was slight. Moreover, because at the time we had so few years over which to detect a population-level change, Patrick didn't even test for one.

As years passed, we captured, banded, recorded, and later recaptured and recorded hundreds more Cassia Crossbills. The opportunity to test further for call divergence presented itself when one of my graduate students, Cody Porter, had the desire and time. Cody's dissertation research focused on the ecology and behavior of reproductive isolation between Ponderosa Pine and Lodgepole Pine ecotypes in the Rocky Mountains. But he had a lull in fieldwork when there was a regional cone failure, with few crossbills to study within hundreds of miles of Laramie. Cody had helped with the banding of Cassia Crossbills for three summers, so he was familiar with the system and data, and he had a discerning ear and an intense interest in vocalizations and evolution.

Like Patrick, Cody found that the contact calls of individual Cassia Crossbills diverged over time from those of the Ponderosa Pine ecotype (Porter and Benkman 2019). Figure 6.4 shows spectrograms of one such Cassia Crossbill recorded in 2013 and 2014. By 2014, its initial component became strongly ascending, which acts to reduce its similarity to that of the Ponderosa Pine ecotype (on the right of Fig. 6.4). Measures of call similarity range from zero, where the calls lack any similarity, to one, where the calls are identical. The average similarity values for the call of this Cassia Crossbill to those of multiple individuals of the Ponderosa Pine ecotype to which Cody made comparisons decreased on average from 0.36 in 2013 to 0.06 in 2014 (Fig. 6.4). This is a particularly striking example. On average, the similarities between the calls of 801 individual Cassia Crossbills recorded over multiple years decreased by 0.0123 per year relative to those of the Ponderosa Pine ecotype.

Although an annual change of 0.0123 seems small, these shifts add up if they are transmitted to their offspring. Indeed, they do add up! Figure 6.5 shows how the population-level similarity between the contact calls of Cassia Crossbills and the Ponderosa Pine ecotype decreased between 1998 and 2018. The average annual decrease in similarity was 0.0047, which is 38% of the annual decrease found within individuals. We would expect the population-level decrease to approximate the individual-level decrease if young crossbills imitated the calls of their parents perfectly. But they don't. We don't know the heritability of vocalizations, but the observed difference in the two rates would be expected if heritability of calls was 0.38. This value is similar to the average heritability of 0.35 for communicative

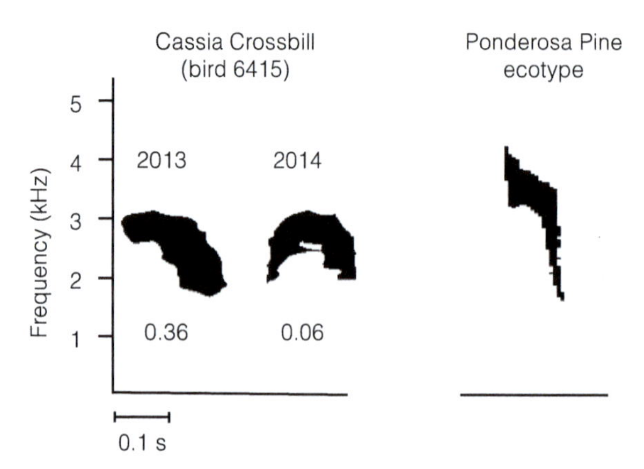

FIGURE 6.4 Spectrograms of contact calls from one Cassia Crossbill *Loxia sinesciuris* captured in 2013 and then in 2014 on the left, and on the right a representative Ponderosa Pine ecotype. The values under the Cassia Crossbill spectrograms represent the average similarity to a sample of calls from the Ponderosa Pine ecotype where one represents identical calls and zero represents no similarity. From Porter and Benkman (2019).

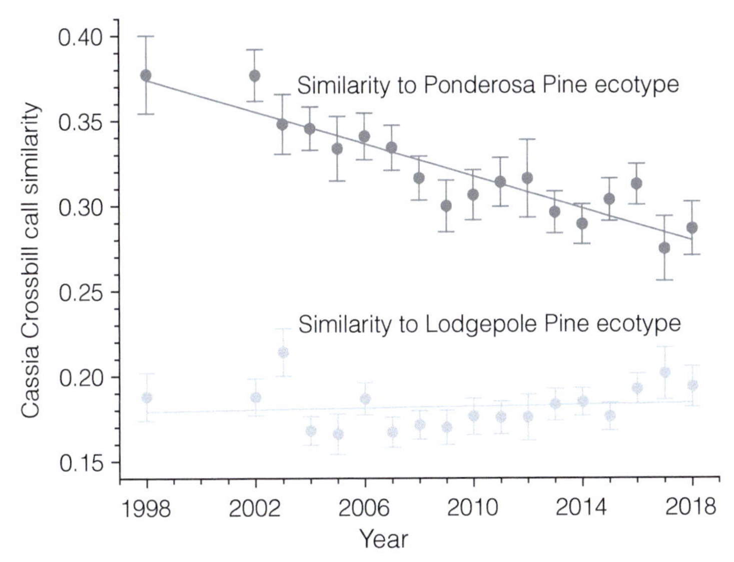

FIGURE 6.5 The average population-level call similarity between Cassia Crossbills *Loxia sinesciuris* and the Ponderosa Pine ecotype decreased over time, while that between Cassia Crossbills and the Lodgepole Pine ecotype remained low and stable (n = 3,242 Cassia Crossbills and 16 of both the Ponderosa Pine and Lodgepole Pine ecotypes). Using one set of contact calls from both the Ponderosa Pine ecotype and the Lodgepole Pine ecotype for comparisons was justified because Cody did not detect trends in their calls over the same time period. The data points represent the mean correlation values and the 95% confidence interval. Modified from Porter and Benkman (2019).

behaviors in animals (based on four studies; Dochtermann et al. 2019). Regardless, if calls change in a consistent direction in adult individuals and such changes are passed on to their offspring because offspring imitate their parents, then changes at the population level over time as in Figure 6.5 are expected. These changes represent cultural evolution—that is, behaviors modified because of social interactions that are passed down to offspring. In Cassia Crossbills, adverse interactions by flocking with Ponderosa Pine ecotypes—resulting in poor patch assessment—would favor call divergence to aid in ecotype discrimination.

Cody also measured call similarity to a sample of the Lodgepole Pine ecotype, which are the second most common ecotype within the range of Cassia Crossbills. The Lodgepole Pine ecotype has calls that are more distinctive from those of Cassia Crossbills (Fig. 6.6) and the call similarity between the two has not changed over time (Fig. 6.5). Thus, it appears that Cassia Crossbill calls have diverged from those of the more similar-sounding (and more commonly co-occurring) Ponderosa Pine ecotype without compromising their already high level of divergence from the Lodgepole Pine ecotype. Whether changes in Cassia Crossbill calls were constrained by their similarity to those of the Lodgepole Pine ecotype is unknown. However, it is conceivable that most changes that would enhance differences from the

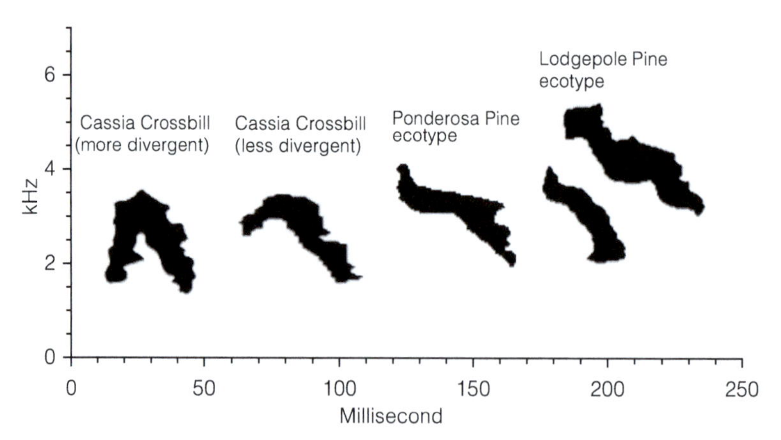

FIGURE 6.6 Sample spectrograms of four contact calls pasted on a common scale: two Cassia Crossbills *Loxia sinesciuris*, and one each from the Ponderosa Pine and Lodgepole Pine ecotypes. The calls from the Cassia Crossbill represent one that is more divergent and one that is less divergent from those of the Ponderosa Pine ecotype. From Porter and Benkman (2019).

Ponderosa Pine ecotype (e.g., the extent of the first ascending element of the Cassia Crossbill call; Porter and Benkman 2019) might not affect their similarity to calls of the Lodgepole Pine ecotype.

Although the decrease in call similarity over time is striking (Fig. 6.5), the key question is whether this divergence in Cassia Crossbill calls affects their ability (and that of the Ponderosa Pine ecotype) to discriminate between their calls and thus facilitate their propensity to associate and flock assortatively. In particular, are Cassia Crossbills more likely to associate with individuals who produce the more divergent calls, and are crossbills of the Ponderosa Pine ecotype better able to discriminate and less likely to associate with such individuals? If the answers are both yes, then it is plausible that such cultural evolution in the Cassia Crossbill has been favored so as to reduce heterospecific flocking. Cody answered these questions by using playbacks of calls from Cassia Crossbills, much like Julie had done earlier using recordings from multiple call types. To select calls for playbacks, Cody ranked 346 Cassia Crossbills captured and recorded in 2016 in terms of call similarity to the Ponderosa Pine ecotype. He then chose recordings from 10 individuals from the upper quartile value of call similarity to the Ponderosa Pine ecotype to represent the less divergent, high-similarity playbacks. The upper quartile value occurs where a total 25% of the individuals had higher similarity values and 75% of the individuals had lower similarity values. To represent the more divergent, low-similarity playbacks, he chose 10 individuals from the lower quartile value of call similarity. The lower quartile value is where 75% of the individuals had higher similarity values.

Cody conducted the playbacks in May and June 2018 when we were banding crossbills in the South Hills. Cody would help with banding early in the morning when we were busiest, often capturing 20 or more crossbills, but then drove off to conduct playbacks in mid-morning and then later in the afternoon. It was a good year to conduct the experiment, because crossbills of the Ponderosa Pine ecotype were common. Cody randomized the playbacks so that he did not know whether he was playing more or less divergent calls at any given time. He'd play only one or the other during any given morning or afternoon. Although Cody didn't know for sure which he was playing, he thought he at least occasionally knew while listening to the playbacks. The afternoon before Cody was set to finish, he was discouraged, believing that Cassia Crossbills had not been more likely and individuals of the Ponderosa Pine ecotype less likely to land in response to the more divergent calls—in other words, that our hypothesis had not been supported. He asked that I quickly analyze the results to see if this discouragement was justified. Cody gave me two files: one with the data, and one with the key to assign each trial to a call similarity value. I was asked to run the analyses because he didn't want to see the key and thereby know the order of playbacks for the next day.

Instead, we were surprised and ecstatic by the pattern in the results. Cassia Crossbills were over 1.6 times more likely to land in response to the more divergent calls (low call similarity to the Ponderosa Pine ecotype calls) than to less divergent calls (Fig. 6.7A), whereas individuals of the Ponderosa Pine ecotype were 60% less likely to land in response to the more divergent calls (Fig. 6.7B). Thus, not only had the calls of the Cassia Crossbill diverged from those of the Ponderosa Pine ecotype, but this divergence had aided these crossbills in discriminating between each other, presumably to facilitate and to accrue the benefits of accurate recognition (ten Cate and Rowe 2007; Grant and Grant 2010). It is the only time that I can truly say I have 'watched' adaptive evolution in action, and it may help explain why

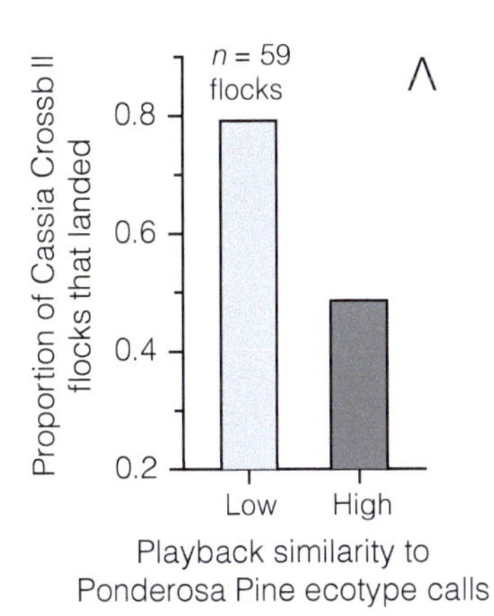

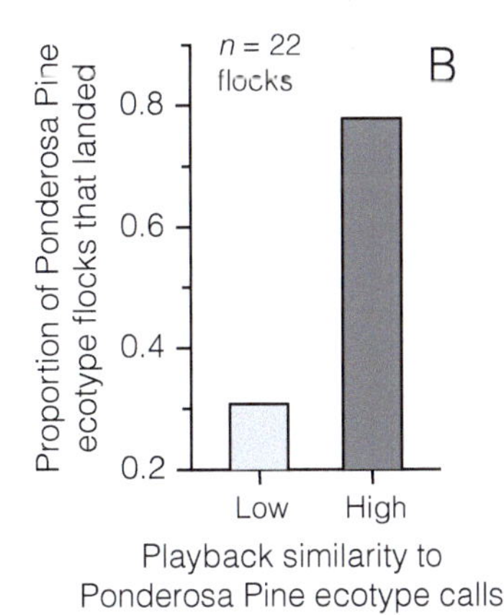

FIGURE 6.7 Flocks of Cassia Crossbills *Loxia sinesciuris* were more likely to land (A) and flocks of the Ponderosa Pine ecotype were less likely to land (B) in response to more divergent (low-similarity; light gray bars) Cassia Crossbill contact calls, supporting the hypothesis that ongoing call divergence in the Cassia Crossbill enhances assortative flocking. From Porter and Benkman (2019).

I seem to have gotten better over the decades at distinguishing Cassia Crossbills, even as my hearing has declined.

Our results suggest why contact calls have proven useful for categorizing crossbills. Contact calls identify helpful flockmates for tree or patch assessment and help coordinate flocking behavior. Furthermore, the often-abundant conifer seeds—the quantity of which varies largely independently of consumption by crossbills—reduces the potential for food competition and thereby provides conditions where the resource assessment advantages we have found for crossbills have the potential to be a primary benefit of flocking (Gil et al. 2017). Flocking has tremendous consequences because crossbills flock year-round and appear to choose mates from within flocks. Indeed, the potential consequences of assortative flocking on mate choice was our main motivation for studying the occurrence and advantages of assortative flocking. It might not be a surprise, therefore, that in the next two chapters I consider how and when assortative flocking along with other factors contribute to reproductive isolation between ecotypes.

Chapter 7

Speciation in the Cassia Crossbill

> Those forms which possess in some considerable degree the character of species, but which are so closely similar to some other forms, or are so closely linked to them by intermediate gradations, that naturalists do not like to rank them as distinct species, are in several respects the most important to us.
>
> (C. Darwin 1859: 47)

Although not specifically referencing crossbills, Darwin described well the near-continuous size variation linking ecotypes and most taxa of Red Crossbills *Loxia curvirostra*, including those forms recognized as distinct species (e.g., Parrot *L. pytyopsittacus*, Scottish *L. scotia* and Cassia *L. sinesciuris*). Darwin also anticipated the differing opinions of ornithologists concerning this variation (reviewed by Groth 1993a). Initially, much of the diversity of Red Crossbills was described as variation among subspecies (Griscom 1937), several of which were used as exemplars of divergence in geographic isolation (allopatric divergence; Mayr 1947). Mayr (1947) inferred that the breeding ranges of the different crossbill subspecies rarely overlap—which is integral to the designation of subspecies—and each had evolved in allopatry. This was a reasonable inference for some of the crossbill subspecies but not all, and the frequent co-occurrence (sympatry) of crossbill 'subspecies' during breeding has been increasingly recognized (Knox 1992; Groth 1993a).

With this recognition and with extensive field and museum study, Groth (1988, 1993a,b) argued that most of the variation within North American Red Crossbills does not reflect subspecific variation. Instead, crossbills are more effectively categorized by vocalizations into 'call types' and these different call types represent different sibling species (i.e., 'morphologically similar or identical populations that are reproductively isolated'; Mayr 1963: 671). Although Groth's (1993a) genetic analyses using allozymes were inconclusive, other lines of evidence were consistent with Red Crossbills being comprised of multiple sibling species. For example, Groth (1988, 1993a) showed that the range of size variation within each call type was similar to that found in other species of songbirds. He reasoned that this implied reproductive isolation between the call types, otherwise, if they interbred freely, the variation within a call type should be elevated. In addition, Groth (1993b) inferred complete assortative mating and hence reproductive isolation between call types from the vocalizations of 24 male–female pairs captured at his study site in the Appalachian Mountains in Virginia. Reproductive isolation between sympatrically breeding forms is one of the hallmarks of species (Mayr 1963; Coyne and Orr 2004).

Subsequent genetic studies revealed little genetic variation among call types in North America and among Red Crossbills in Eurasia (Questiau et al. 1999), and even between recognized species (Red, Scottish and Parrot; Piertney et al. 2001). These genetic analyses were standard for the time, and studies of other bird groups using the same genetic markers revealed evidence of independent evolution and reproductive isolation in closely related taxa. However, the analyses by Questiau, Piertney and their colleagues were limited to a tiny fraction of the genome and would not necessarily detect genetic differences if crossbill lineages had diverged recently and population sizes were large, conditions that slow neutral genetic divergence. (Evolutionary biologists use 'neutral' to refer to changes in the genetic composition of populations that are due to random chance, not natural selection.) In fact, it was plausible that many of the call types evolved following the expansion of conifers in the last 9,000–18,000 years, with plenty of opportunity for gene flow (Benkman 1989c, 1993a).

Contemporary data on assortative pairing and reproductive isolation were needed. Groth (1993b) used call matching (of contact calls) of the crossbills he captured at his netting site in Virginia to categorize birds as mated pairs and to infer assortative mating. However, by his definition only assortative pairs (i.e., those with matched calls) were recognized. Thus, if there were mixed pairs they would not have been recognized as mated pairs. In addition, Groth found evidence of breeding for only 3 of the 24 pairs. If after breeding, mixed pairs are less likely to associate than assortative pairs, then samples of nonbreeding pairs would be biased further toward being assortative. What we needed were studies that focused on finding breeding pairs and then recording their calls to determine if breeding pairs were assortative. However, reliably finding lots of breeding pairs of nomadic ecotypes year after year poses a formidable challenge.

Cassia Crossbills and some natural history

After capturing and recording what eventually would be recognized as Cassia Crossbills, I realized that I had found a remarkable system for study (Chapter 5, 'Finding another Cypress Hills'). Cassia Crossbills are resident in the South Hills/Albions because of the exceptional stability of their seed resource. Over 90% of the Lodgepole Pine in these two ranges produce serotinous cones that remain attached to branches and closed for a decade or more (Fig. 7.1). In addition,

FIGURE 7.1 Photograph of multiple years of serotinous Rocky Mountain Lodgepole Pine *Pinus contorta* var. *latifolia* cones along a branch, with increasingly older cones toward the back. Each whorl of cones approximates a year of cones; there are about eight years of cones along the main branch.

seed production is remarkably consistent from year to year (Fig. 7.2; Box 2.1). Thus, large numbers of cones accumulate in the canopy at a roughly constant rate. Once the cones mature, dry and harden in the fall they are impenetrable to crossbills. It is only after the cones weather—usually for seven or more years—that crossbills can access seeds between slight gaps that form between some of the scales in some of the cones (Fig. 5.6; see Fig. 7.8B). During a given morning, a crossbill might remove seeds from cones that vary in age by one or more decades. Because of gradual weathering of serotinous cones that accumulate yearly in the canopy, seeds become available throughout the year, year after year, allowing

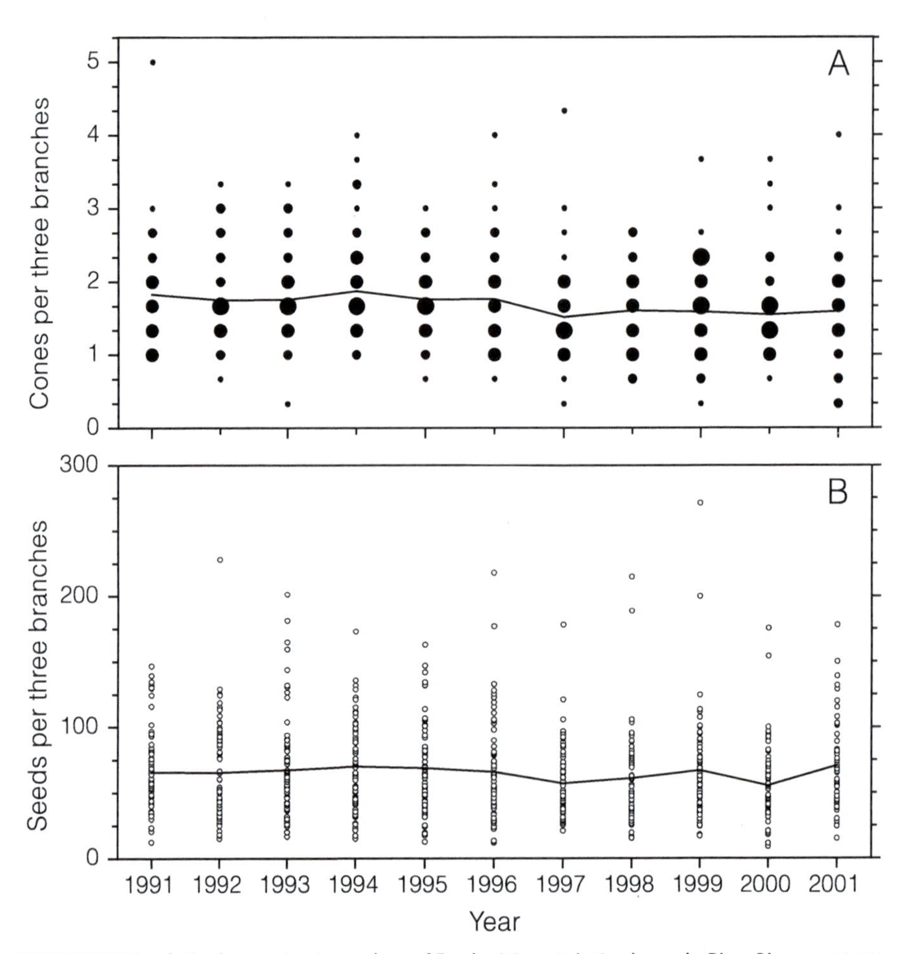

FIGURE 7.2 A relatively constant number of Rocky Mountain Lodgepole Pine *Pinus contorta* var. *latifolia* (A) cones and (B) full seeds is produced each year, as indicated by the similarity in the average number of cones and seeds produced annually on three upper branches from 78 mature Lodgepole Pines in the South Hills, Idaho. The lines connect the yearly means. The size of the circles in (A) increase with increasing sample size (1–2, 3–10, 11–20 and >20 trees). Modified from Benkman et al. (2012).

both residency and high densities of crossbills (Benkman 1999; Siepielski and Benkman 2005; Benkman et al. 2012). And thus, the potential for us finding lots of breeding pairs year after year.

Reproductive isolation of the Cassia Crossbill

Soon after I returned to New Mexico after capturing and recording Cassia Crossbills, I spoke to Julie Smith about the opportunity to study Cassia Crossbills. Julie was finishing her master's thesis on the benefits of assortative flocking, which you learned about in the previous chapter. Cassia Crossbills were particularly suitable for studying reproductive isolation because they likely evolved recently, perhaps in the last 5,000 years (Chapter 5, 'A geographic mosaic of coevolution'). Lineages in the early stages of divergence are especially useful for study because the causes of speciation are more likely to be evident and active (Coyne and Orr 2004). Thus, if Julie could attribute reproductive isolation to particular aspects of their biology (termed 'reproductive isolating barriers'), we would be on firmer ground inferring the mechanisms for the evolution of reproductive isolation. In contrast, species that diverged hundreds of thousands of years or longer ago might now be reproductively isolated for reasons having little to do with the initial causes of divergence.

Julie began searching for breeding pairs and recording their contact calls on 1 February 2001 and remained in the field until near the end of July, when nesting was no longer initiated. Breeding pairs were recognized by behaviors associated with nesting, including courtship feeding and nest building (Knox 1990; Benkman and Young 2020). Julie returned to the South Hills in February 2002 and remained until the beginning of August. By then it became clear that most Cassia Crossbills initiated nests in April and continued through June, with few nests initiated after July. Nesting was quite seasonal and occurred in the same area year after year, more like a standard songbird than the vast majority of crossbills (Chapter 2, 'Timing of breeding'). We assumed that social pairs reflected mating pairs and that extrapair paternity, if it occurred, did not alter our measures of assortative mating (Smith and Benkman 2007; Porter and Benkman 2022; I address this assumption in Box 7.1).

Julie—who, as by now you have probably gathered, is a talented and indefatigable biologist—found 428 breeding pairs, for which she was able to record the contact calls of both mates. In four subsequent years (2003–6), Lenny Santisteban found and recorded 447 pairs, for a total of 875 pairs over the six years (2001–6). A total 1,704 or 97.4% of the 1,750 crossbills found in mated pairs were Cassia Crossbills. The remaining 46 individuals were of the Ponderosa Pine and Lodgepole Pine ecotypes: 30 and 16, respectively. Only 12 Cassia Crossbills were in mixed pairs—that is, 0.7% of the breeding Cassia Crossbill population were in hybrid pairs. No hybrid pairs were found between the Ponderosa Pine and Lodgepole Pine ecotypes. (However, this reveals little because of the rarity of the two pine

Box 7.1: Rarity of extrapair paternity

Individuals of many bird species engage in extrapair copulations that result in extrapair paternity (Brouwer and Griffith 2019). If extrapair paternity occurred in crossbills such that it altered estimates of assortative pairing, then this would compromise our study. However, for reasons that I explain below, I believe extrapair paternity is rare in Red Crossbills and unlikely to bias our results (Smith and Benkman 2007; Porter and Benkman 2022). The most direct evidence supporting this assumption is from Kleven et al. (2008) who found no evidence for extrapair paternity among 96 offspring from 34 Red Crossbill broods in Norway. I believe that the absence of extrapair paternity is universal to other populations within the Red Crossbill complex because their breeding behavior is widely similar (Cramp and Perrins 1994; Benkman and Young 2020). Extrapair paternity is likely rare if not absent because male Red Crossbills vigorously mate guard females until egg laying (Nethersole-Thompson 1975; Benkman and Young 2020). In addition, female crossbills rely exclusively on their social mate for food for themselves and for their young when they are brooding their nestlings during the first five days after hatching (Chapter 2, 'Timing of breeding'). In bird species where male parental care is critical to female reproductive success, extrapair paternity is rare or absent (Møller 2000). Moreover, little to no extrapair paternity is consistent with the relatively small testes of the Ponderosa Pine and Lodgepole Pine ecotypes (Smith and Benkman 2007), which are comparable in relative size to those of other genetically monogamous bird species (Møller and Briskie 1995). Finally, the relatively short sperm and elevated levels of sperm length variation within and among male Red Crossbills (Lifjeld et al. 2010) are all consistent with little to no extrapair paternity (Lifjeld et al. 2010; Birkhead and Montgomerie 2020). Of the 55 bird species surveyed by Lifjeld et al. (2010), only one had greater sperm length variation among males than Red Crossbills.

Furthermore, some extrapair paternity need not alter our measures of assortative pairing, unless individuals in for example mixed pairs seek extrapair copulations with individuals of their own ecotype. Few data are available on extrapair mating and interbreeding between lineages, but the available evidence suggests that social pairing behavior is strongly correlated with extrapair mating behavior in young, recently diverged bird lineages like crossbills (Vallender et al. 2007; Turbek et al. 2021). Cases of extrapair mating deviating from social pairing with respect to lineage identity have been documented in systems with severe costs of hybridization (i.e., complete intrinsic postzygotic isolation, such as occurs when hybrid offspring are sterile; Veen et al. 2001) or where males of one lineage are socially dominant to males of another (Reudink et al. 2006), neither of which characterizes the crossbills we studied.

ecotypes.) Although Julie found pairs in the Albion Mountains, the vast majority of our research has been in the South Hills where Lodgepole Pine is more extensive and accessible. Hereafter, for simplicity I usually refer to just the South Hills when discussing the Cassia Crossbill.

At this point it is worth mentioning that our estimate of the frequency of hybridization is likely to be a slight underestimate. This is because, on very rare occasions, Cassia Crossbills gave a categorically different call in a subsequent year. For example, 3 of 844 birds initially giving Cassia Crossbill calls later gave a call that Cody Porter categorized as the Ponderosa Pine ecotype (Porter and Benkman 2019). In one case, a female Ponderosa Pine ecotype paired with a male Cassia modified her call to match that of her mate during their second year together (Keenan and Benkman 2008). Thus, some pairs that we categorized as assortative by their contact calls might have represented hybrid pairs where one partner changed its calls. Based on this information, we estimate that the occasional call switching causes us to underestimate the occurrence of hybridization by 0.36% (Benkman et al. 2022). Our estimate of 0.7% of breeding Cassia Crossbills in hybrid pairs therefore increases to just over 1%.

Call switching aside, the most obvious question is why so few of the pine ecotypes bred in the South Hills. It is not because geographic barriers prevent dispersal to the South Hills. Although the South Hills are on the edge of the distribution of Rocky Mountain Lodgepole Pine *Pinus contorta* var. *latifolia*, isolated by sagebrush steppe from mountains to the north and east (see Figs 5.2 and 5.5), the South Hills are not particularly isolated for a crossbill. Every year some to many of both pine ecotypes arrive in the South Hills, especially during the two main periods of movement: late spring to early summer, and late fall (Chapter 2, 'Crossbill responses'). During summer, pine ecotypes can represent up to 20% of the crossbills in the South Hills. Yet few remain through fall, winter and spring. In similarly isolated ranges with Lodgepole Pine and lacking Red Squirrels, but without an endemic, locally adapted crossbill (e.g., the Little Rocky Mountains in Montana, east of the Rocky Mountains), the Ponderosa Pine ecotype occurs at high densities (Siepielski and Benkman 2005).

If geographic isolation cannot account for why few individuals of the pine ecotypes remain in the South Hills, then what explains their rarity? The answer appears to be competition. Not interference competition (agonistic behavior)—crossbills are not territorial—but rather *indirect* competition. Cassia Crossbills appear to reduce the availability of seeds, making it difficult for non-locally adapted crossbills to survive through fall and winter (Chapter 3, 'Performance-based habitat choice'). I infer such exploitative competition based on analyses of intake rate data that Julie and Trevor Fetz gathered.

I had anticipated that Cassia Crossbill intake rates in the South Hills would be fairly stable year-round, because of what I suspect is a nearly constant rate of gap formation between scales as cones of various ages weathered from variable exposure to sun and moisture. However, Julie and Trevor did not find stable intake rates over two years of data gathering. Instead, intake rates increased from late fall to early spring and then declined in a very seasonal manner (Fig. 7.3). Such a pattern might arise from a pulse of gap formation between the scales beginning in late fall and continuing through winter, but I cannot explain why this would occur in the first place. If there *were* such a pulse in gap formation, I would expect it in summer when temperatures are hottest (Chapter 9, 'Climate change, Cassia Crossbills and early seed shedding'). An alternative was that seasonal variation in the depletion of seeds by crossbills caused seasonal variation in intake rates.

To test this idea, we enlisted the help of a graduate student, Matt Talluto. Matt's primary focus was on the variation in the frequency of serotiny in Lodgepole Pine in relation to variation in fire frequency and seed predation by Red Squirrels, described in Box 5.1. Luckily, Matt's analytical and programming skills allowed him to mathematically model whether seasonal seed demand could drive the observed intake rate dynamics (Benkman et al. 2012). He first estimated the daily energy demand of a population of Cassia Crossbills throughout the year, using data for crossbills and a bioenergetic model developed for birds (Wiens and Innis 1974; Rexstad 1982). Assuming that new seeds become available at a constant rate, Matt modeled intake rates as seed depletion varied with seasonal demand. His predicted intake rates mirrored rather well those observed (the dashed curve as compared to the solid curve, respectively, in Figure 7.3). Intake rates begin to peak just before the large increase in energy demands (and associated seed depletion) as eggs hatch, and then intake rates decline from the accelerating energy demand of additional mouths to feed after young are hatched. Caring for nestlings is the period of peak energy demand for most songbirds (Walsberg 1983). After the young are fledged and become independent of their parents, mortality increases for the offspring (as is common in songbirds; Sullivan 1989; Naef-Daenzer and Grüebler 2016) and overall energy demand declines. Intake rates then increase as seed renewal exceeds demand from fall until the eggs begin to hatch the following April.

These results indicate that seed depletion by Cassia Crossbills cause declines in seed intake rates and that the interaction between renewal of accessible seeds and depletion by crossbills drives the seasonal dynamics in intake rates (Fig 7.3). Seed depletion by crossbills presumably limits both the length of the breeding season and crossbill survival, especially in late fall and winter. If crossbills increased in abundance, then seed depletion and the resulting depression in intake rates would be even greater, causing the crossbill population to decline. If crossbills decreased in abundance, then intake rates would increase allowing higher survival, more reproduction and a rebound in the crossbill population. Thus, population density is regulated by the balance of seed renewal and depletion (Benkman et al. 2012). This is highly unusual for crossbills because most often variation in seed intake rates appears related to cone ripening phenology (Chapter 2). Intake rates normally increase as seeds mature and cones open, and then decline as seeds are shed. With

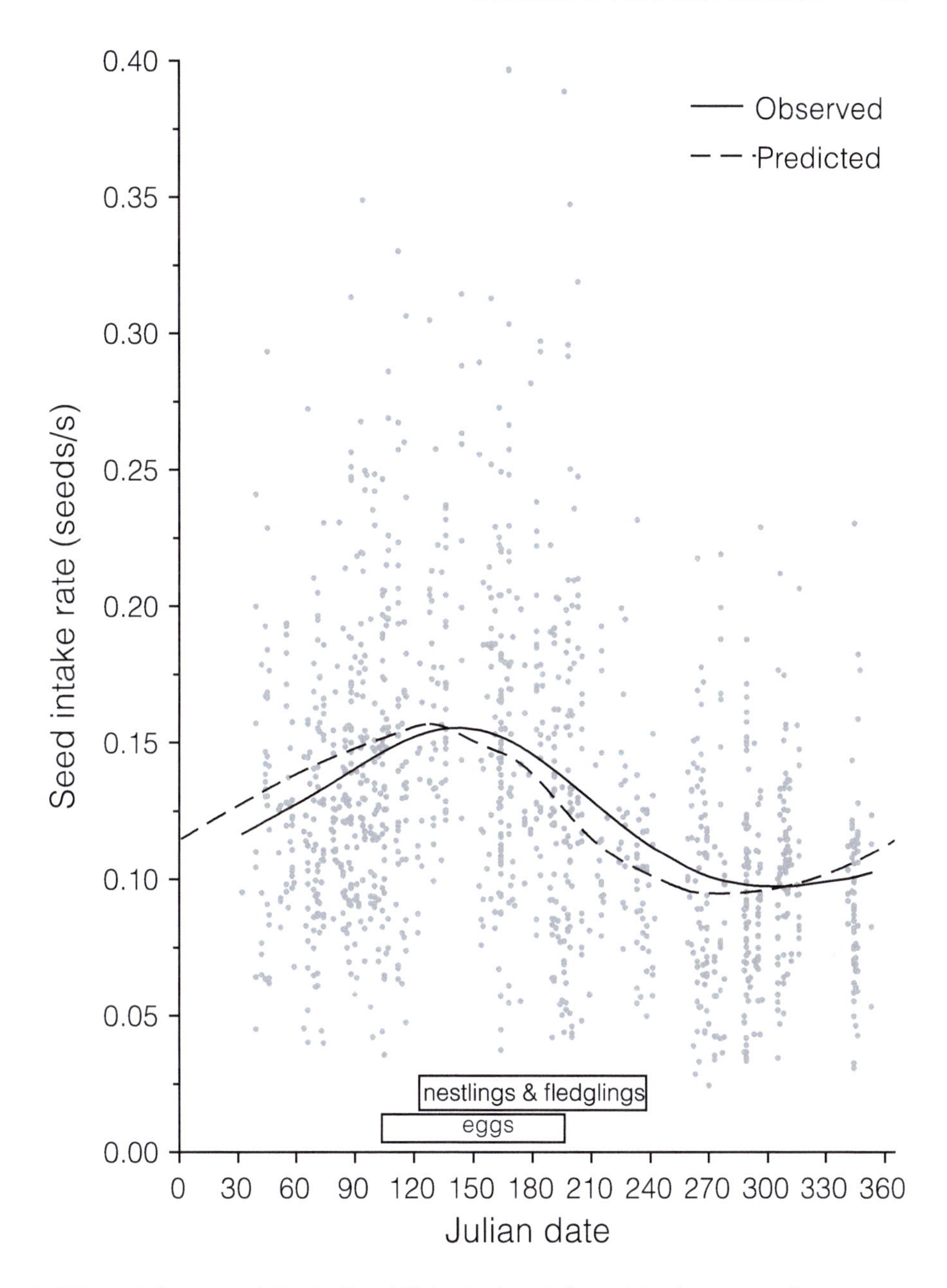

FIGURE 7.3 I expected Cassia Crossbill *Loxia sinesciuris* seed intake rates to be constant throughout the year; however, intake rates varied seasonally (solid curve fit to 1,266 Cassia Crossbill foraging bouts [gray points] using a cubic spline), as would be predicted if the constant replenishment of seeds was depleted in proportion to total estimated energy demands (dashed curve). Modified from Benkman et al. (2012).

the usual large fluctuations in seed crops, crossbills have a limited impact on seed intake rates except during the occasions when crossbill densities are high relative to seed abundance (Benkman 2020).

Our results imply that Lodgepole Pine seeds are a limiting resource for Cassia Crossbills, especially in fall and winter when intake rates are low (Fig. 7.3), and day length and temperature are decreasing. I believe this prevents the two pine ecotypes, which on average are less well adapted for foraging on seeds in the cones in the South Hills, from remaining through winter (Fig. 3.7; Benkman 2017a). Consequently, pine ecotypes are scarce when Cassia Crossbills begin courting and pairing in January and February. Although intake rates are relatively high in late March and April when Cassia Crossbills begin nesting (Fig. 7.3), few individuals of the pine ecotypes are present. The pine ecotypes tend to disperse into the South Hills mostly in May and June, after most Cassia Crossbills are paired and nesting. In addition, intake rates at this time are declining and not particularly high compared to those when feeding on developing Engelmann Spruce *Picea engelmannii* and Ponderosa Pine *Pinus ponderosa* cones later in the summer, when the pine ecotypes commonly nest (Porter and Benkman 2022). This is presumably why most individuals of the pine ecotypes seemingly move on from the South Hills (Smith and Benkman 2007). The result is that competitive exclusion of pine ecotypes by Cassia Crossbills spurs habitat isolation, limiting the opportunity for interbreeding. This is similar to the 'monopolization hypothesis' proposed by De Meester and colleagues (2002) to account for a reduction in interbreeding between locally adapted populations and maladapted immigrants.

Habitat isolation is a widely recognized reproductive isolating barrier that arises as populations adapt to alternative resources and habitats, such that they are unlikely to breed in the same habitats (Coyne and Orr 2004). But habitat isolation is not the only barrier to interbreeding. Not only do relatively few pine ecotypes remain in the South Hills, but those that remain are less likely than Cassia Crossbills to breed. About 30% fewer of the pine ecotypes in the South Hills attempted to breed compared to Cassia Crossbills. The result is that there are even fewer opportunities for interbreeding than if habitat isolation were the only barrier. Much as for habitat isolation, we attribute the lower propensity for breeding by pine ecotypes to them being less efficient on average than Cassia Crossbills at foraging on the cones in the South Hills, with the result that they struggle to reach prime reproductive condition (Smith and Benkman 2007).

This lower propensity for breeding when ill-adapted to a particular habitat is similar to immigrant inviability, which is the inability to survive in the habitat where another taxon prospers (Nosil et al. 2005). In many plants and invertebrates, for example, abilities to move and choose among habitats are limited and individuals are more likely to succumb if they end up in unsuitable habitats. However, as discussed in Chapters 2 and 3, crossbills have superb abilities to both assess resource or habitat quality and move among habitats, spending little time in those locations where they would do poorly and remaining in those where they thrive (performance-based habitat choice; Chapter 3). Thus, we suspect crossbills and perhaps most birds move on from ill-suited habitats rather than remain and die, especially during generally favorable periods when they are likely to find areas where they can breed.

Finally, the crossbills that breed in the South Hills do so assortatively. That is, given the relative abundance of breeding Cassia Crossbills and pine ecotypes, the frequency of interbreeding is much lower than expected if pairing was random (Smith and Benkman 2007). This is termed 'behavioral isolation' and is probably the result of multiple factors. First, the songs sung by male Cassia Crossbills differ from those sung by the pine ecotypes (Porter and Smith 2020). We have not tested for female preference of song, but we suspect differences in the songs between the different forms are important in assortative pairing, as in other birds (Grant and Grant 2008).

Another factor contributing to behavioral isolation is a preference for associating with individuals producing the same contact calls as their own. I was fortunate to be able to recruit Lisa Snowberg to test for such preferences for her master's thesis. Lisa had spent the previous year in Geoff Hill's lab at Auburn University conducting mate choice experiments in birds, so she was already very knowledgeable about the methodologies and conceptual foundation of mate choice studies. Using Cassia Crossbills and individuals of the Ponderosa Pine ecotype, Lisa tested whether females stimulated into reproductive condition with estradiol preferentially associate with their own type when provided a choice between one male of each type that were of similar size and coloration. The males were not singing, and contact calls were the one obvious difference between them (the males could not see each other or interact physically because they were separated by a solid barrier). Lisa found that females of both types preferentially associated with males of their own type (Snowberg and Benkman 2007). Because the females were in breeding condition, this suggests that vocalizations other than song contribute to assortative mating.

Earlier, I mentioned that in comparison to Cassia Crossbills, pine ecotypes were about 30% less likely to form pairs and breed, because they were less efficient and thus less likely to attempt to reproduce. However, this might also reflect the inability of relatively inefficient foragers to attract a mate, a scenario which falls under the umbrella of behavioral isolation. In a second set of experiments, Lisa tested whether crossbills might preferentially pair with faster foragers by presenting female Cassia Crossbills with the choice of two similarly sized and colored male Cassia Crossbills that differed in feeding performance. Lisa altered feeding intake rates by opening serotinous Lodgepole Pine cones in an oven and removing seeds from one set of cones while leaving the seeds in those that remained. She then moistened the cones to reclose them. In each trial, a satiated but estradiol-stimulated female was allowed to choose between two hungry males vigorously foraging on two sides of a solid partition: one foraging on cones with a full complement of seeds and the other foraging on cones from which three-quarters to five-sixths of the seeds had been removed. Females consistently associated with the male who seemingly was the more efficient forager—that is, required about half the time to remove a seed from the cone (Snowberg and Benkman 2009).

The survival and reproductive benefits to females choosing to mate with a more efficient or locally adapted male are obvious, given her reliance on the male to provide food for both her and her offspring (Chapter 2, 'Timing of breeding'). This acts like condition-dependent mate choice (Edelaar 2022), which accelerates local adaptation (Veen and Otto 2015). Locally adapted males are not only more likely to survive but also more likely to mate and leave offspring, which could reduce gene flow between populations by preventing mating with immigrants and thereby facilitate speciation (van Doorn et al. 2009; Veen and Otto 2015).

A final contributing factor to assortative pairing and behavioral isolation is the propensity of crossbills to assortatively flock with individuals giving the same contact calls, as discussed in the previous chapter. Given that assortative flocking is quite strong among crossbills in the South Hills (Porter and Benkman 2019) and that Cassia Crossbills generally pair in late winter while in flocks (Smith and Benkman 2007), assortative flocking likely contributes to assortative pairing. I revisit the importance of assortative flocking in Chapter 8.

We now had data on a number of reproductive barriers that might be contributing to speciation among Cassia Crossbills and other ecotypes. How do they add up? We used metrics developed by Coyne and Orr (1989) and modified by Ramsey et al. (2003) to estimate the level of reproductive isolation attributable to habitat isolation, the lower propensity of pine ecotypes to attempt breeding, and behavioral isolation. A value of zero represents random mating, while a value of 1 represents complete reproductive isolation between populations. We found that on this scale, the level of reproductive isolation between Cassia Crossbills and the Ponderosa Pine ecotype was 0.998, and even closer to 1 between the Cassia Crossbill and the Lodgepole Pine ecotype (Smith and Benkman 2007). These figures ignore both call switching and the reproductive isolation that might occur if hybrid individuals are less likely to survive and reproduce than the parental forms, though such postzygotic reproductive isolation (meaning reproductive isolation after the egg is fertilized) is likely also at play in crossbills (Snowberg and Benkman 2007). Our results therefore indicate very strong reproductive isolation between Cassia Crossbills and their congeners (Benkman et al. 2022). The next question was whether such high levels of reproductive isolation also occurred in the past, and not just over the six years of study.

To address this, we needed to analyze a much greater proportion of the genome than had been done in earlier studies, looking for evidence that limited mating between residents and nomadic ecotypes had allowed genetic differences to accumulate. Though we were able to use some of the blood samples from Cassia Crossbills and the pine ecotypes that we had banded, we also relied extensively on frozen tissues from specimens of all the call types that Jeff Groth had collected for his dissertation (Groth 1993a). The specimens and associated tissues that Jeff collected are preserved and curated at the Museum of Vertebrate Zoology at UC

Berkeley, an invaluable resource. The curators of this collection—the late Ned Johnson, who was Jeff's PhD advisor and from whom I took Ornithology while I was an undergraduate, and Carla Cicero—kindly provided us access to them.

Tom Parchman took the lead on this project. As a graduate student, Tom focused mostly on the coevolutionary interactions that I discussed in Chapter 5. However, Tom had become increasingly interested and proficient at using genetic tools to answer evolutionary questions—work that is now his bread and butter as a faculty member at the University of Nevada at Reno. The most striking finding of the resulting study was that the Cassia Crossbill is particularly distinct genetically compared to other types of Red Crossbill in the New World (Parchman et al. 2016). This result was consistent with both the high levels of contemporary reproductive isolation and the strong reproductive isolation in the past. Our genetic analyses in combination with the results of our fieldwork convinced the Classification and Nomenclature Committee of the American Ornithologists' Union (now the American Ornithological Society) to designate the Cassia Crossbill as a distinct species (Chesser et al. 2017). Though not everyone agrees with this designation (Hill and Powers 2021), dissenting opinions have not convinced the Classification and Nomenclature Committee of the American Ornithological Society to change their mind (Chesser et al. 2022; see Benkman et al. 2022).

The coevolutionary arms race between Cassia Crossbills and Lodgepole Pine that causes divergent selection and speciation (Smith and Benkman 2007; Parchman et al. 2016) illustrates what is termed 'coevolutionary diversification' (Thompson 2005, 2013; Althoff et al. 2014; Hembry et al. 2014). A recent and particularly elegant study demonstrating coevolutionary diversification showed that reproductive isolation arose between populations of brood-parasitic cuckoos in Australia as they coevolved with alternative host species of birds (Langmore et al. 2024), and that especially strong coevolutionary interactions (where parasitism causes reproductive failure for the host) accelerated cuckoo diversification. Another example of coevolutionary diversification is the classic example of speciation in threespine sticklebacks in lakes in the Gulf Islands of British Columbia, where coevolution between competitors causes morphological divergence (character displacement) that drives reproductive isolation (Schluter and McPhail 1992; Schluter 1996; Rundle et al. 2000; Schluter 2010). Coevolution between competitors in other adaptive radiations likely provides additional examples of coevolutionary diversification (Schluter 2010).

Given the widespread occurrence of coevolution (i.e., most if not all species are engaged in it) and geographic variation in the occurrence and strength of such interactions, it is not far-fetched to argue that coevolution has figured prominently in the diversification of life (Thompson 2005, 2013). Nevertheless, the relative importance of coevolutionary diversification remains unclear (Althoff et al. 2014; Hembry et al. 2014). Even within systems such as feather lice that parasitize and coevolve with their bird hosts (Clayton et al. 2016), host switches alone can rapidly spur reproductive isolation between populations of feather lice (Villa et al. 2019) and such switches are evidently a dominant process in the diversification of at least some feather lice clades (Clayton et al. 2016). Similarly, host switches and specialization on alternative species of conifers appear much more influential than coevolution in the diversification of crossbills (Chapters 4 and 10). It is likely that many crossbill populations coevolve with conifers (Benkman et al. 2010), but the ensuing changes in crossbills are generally slight relative to those following switches onto new species of conifers, as (for example) occurs after the retreat of glaciers and the expansion of conifer ranges (Benkman 1993a).

Chapter 8

Causes and Consequences of Variation in Reproductive Isolation

> [F]urther work in nature is needed, particularly to show that the barrier evolved via divergent selection and is actually involved in reductions in gene flow. Even once divergent selection has been implicated, the specific mechanistic causes of evolution (e.g., for habitat isolation, the fitness trade-offs…) have rarely been addressed.
>
> (P. Nosil 2012: 101)

Why are none of the remaining Red Crossbill *Loxia curvirostra* ecotypes in North America as genetically distinct as the Cassia Crossbill *L. sinesciuris*? Various factors contribute to genetic divergence, which may impact other populations differently than Cassia Crossbills. Occupying different geographic regions—a condition known as allopatry—reduces if not eliminates interbreeding and is undoubtedly critical for speciation in nearly all birds and most other organisms (Mayr 1963; Coyne and Orr 2004; Price 2008). Allopatry is unlikely to account for the genetic distinctiveness of Cassia Crossbills, however, because they are not particularly isolated geographically compared to other ecotypes.

Another potential variable is time. As speciation requires the passage of generations for genetic differences to emerge and accumulate (Coyne and Orr 2004; Price 2008), evidence that the Cassia Crossbill had split from its congeners earlier than divergence within other call types would help explain the level of reproductive isolation we observe. However, we have no evidence for this or for other substantial differences in divergence times among New World Red/Cassia Crossbills, which appear only to have split in the last 20,000 years. If anything, the Cassia Crossbill might be among the *youngest* of the crossbill lineages in North America (Parchman et al. 2016). In contrast, one New World lineage that may be older than the rest is the Central American subspecies *L. c. mesamericana*, which occupies low-latitude pine forests that have likely been more stable during recent glacial cycles than those farther north (Jansson and Dynesius 2002). Unfortunately, we lack genetic data to address this.

A third variable is divergent selection. Genetic divergence decreases with increasing gene flow (arising from interbreeding) but increases with strengthening divergent selection (Hendry et al. 2001). Although I am hesitant to infer too much from the fitness surface in Figure 4.8, divergent selection associated with Lodgepole Pine *Pinus contorta* in the South Hills does not appear to be especially strong compared to that associated with the conifer species that other ecotypes rely on. In addition, the bills of Cassia Crossbills are not particularly distinctive compared to those of the Ponderosa Pine ecotype, suggesting they have not been disproportionately impacted by this evolutionary force. For example, the difference in average bill depth between Cassia Crossbills and the Ponderosa Pine ecotype (~0.3 mm) is comparable to those between other sympatric pairs of ecotypes (Groth 1993a; Benkman et al. 2009; Irwin 2010).

A fourth variable is population size. As population sizes increase, their rate of neutral genetic divergence decreases, because mutations are less likely to change dramatically in frequency by chance events (Piertney et al. 2001; Hudson and Coyne 2002). Cassia Crossbills undoubtedly have and have had a smaller population size than those of the Red Crossbill ecotypes that we compared them to genetically (Parchman et al. 2016). It is plausible that the various ecotypes are all strongly reproductively isolated like the Cassia Crossbill, and that the greater genetic distinctiveness of the Cassia Crossbill is the result of its smaller population size. Nor can we exclude the possibility that small population size combined with strong reproductive isolation is why Cassia Crossbills are so distinct genetically. However, we *can* test whether reproductive isolation is as strong among the Red Crossbill ecotypes as it is for Cassia Crossbills. This is the topic I address in the rest of this chapter.

Reproductive isolation between the pine ecotypes

Cody Porter was the ideal person to carry out a study to determine if Red Crossbill ecotypes were less reproductively isolated than the Cassia Crossbill. We learned a bit about Cody and his work on assortative flocking and call divergence in the Cassia Crossbill in Chapter 6, 'How might new ecotypes evolve?': passionate about both crossbills and evolutionary biology, he has a keen interest in speciation. Moreover, Cody is a talented field ornithologist undeterred by rough terrain and weather. He is not only a sharp observer, but is also the one to ask when uncertain as to the identity of a calling crossbill (or any other bird!).

Based as I was at the University of Wyoming in Laramie following my move in 2004, the logical pair of ecotypes to study were those associated with Lodgepole Pine and Ponderosa Pine *Pinus ponderosa*. These ecotypes provided a useful comparison because they were the only other crossbills to co-occur with and hybridize (rarely) with the Cassia Crossbill. Based on our genetic analyses, they also appear to be closely related (Parchman et al. 2016). Perhaps even more importantly, they breed regularly in the ranges east and west of Laramie, and to the south in the Colorado Rockies. Both Lodgepole and Ponderosa Pine are widespread in this region of the Rocky Mountains, and although Lodgepole Pine generally occurs at higher elevations than Ponderosa Pine, they commonly overlap in distribution, with patches of each species forming a mosaic. Moreover, Engelmann Spruce *Picea engelmannii* is common at higher elevations, where both ecotypes regularly feed on it while nesting. This is an example of the phenomenon discussed in detail in Chapter 4: Different species and ecotypes of crossbills appear to specialize on conifers with the most reliable crops, but can converge to feed and breed on other species (like Engelmann Spruce) on occasions when their seeds are abundant and easily accessible. (Both pine ecotypes also nest while foraging on Rocky Mountain Douglas-fir *Pseudotsuga menziesii* var. *glauca*; however, this occurred infrequently during Cody's study.)

We were particularly excited to compare levels of reproductive isolation between the two ecotypes while feeding on pine (Lodgepole and Ponderosa) versus while feeding on spruce (Fig. 8.1). A prominent hypothesis in the speciation literature is so-called ecological speciation, or speciation driven by divergent selection for

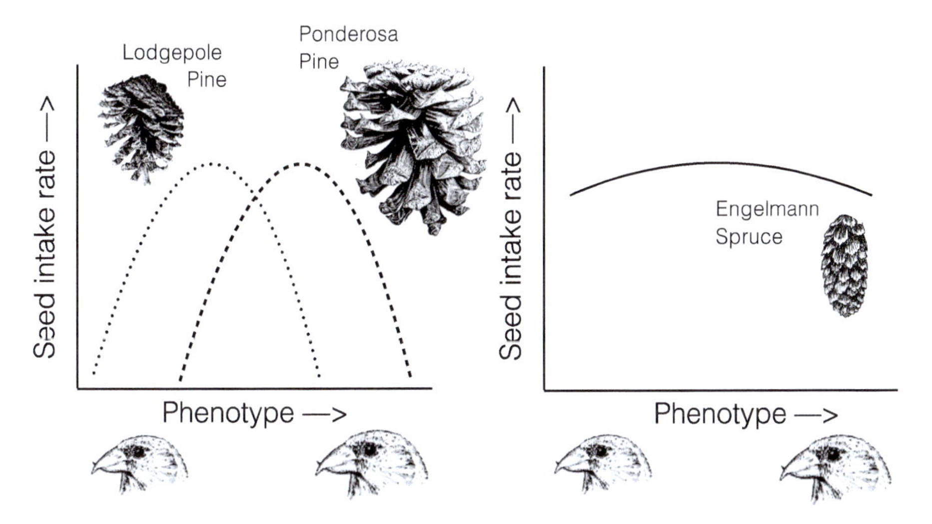

FIGURE 8.1 Both pine ecotypes breed in forests of mixed Lodgepole Pine *Pinus contorta* and Ponderosa Pine *P. ponderosa* (on left), where feeding tradeoffs are substantial (those most efficient foraging on Lodgepole Pine are less efficient on Ponderosa Pine and vice versa; see Fig. 4.8). The most efficient crossbills at foraging on Lodgepole Pine are smaller-billed with narrower husking grooves (dotted curve shows seed intake rates on Lodgepole Pine in relation to bill depth), which correspond to mostly individuals of the Lodgepole Pine ecotype, whereas those most efficient on Ponderosa Pine are larger-billed with wider husking grooves (dashed curve shows seed intake rates on Ponderosa Pine), corresponding to mostly individuals of the Ponderosa Pine ecotype. Both pine ecotypes also breed while feeding on Engelmann Spruce *Picea engelmannii* (on right), where variation in bill structure has less of an influence on seed intake rates and both pine ecotypes have high and similar seed intake rates. Modified from Porter and Benkman (2022).

alternate resource use or habitats that causes reproductive isolation as a byproduct. When this occurs, stronger divergent selection should result in stronger reproductive isolation (Schluter 2001, 2009; Nosil 2012). Considerable evidence has accrued in support of ecological speciation (Nosil 2012), and the Cassia Crossbill is a prime example (Price 2008). However, divergent selection is generally inferred in a qualitative manner (e.g., adaptation or use of two host plant species or two habitat types). Quantitative measures demonstrably related to divergent selection such as seed intake rates for crossbills are rare. Based on data from aviary studies (Chapter 4), strong divergent selection (i.e., strong feeding tradeoffs) occurs when crossbills forage on conifers involved in ecotypic specialization: the Lodgepole Pine ecotype is much more efficient than the Ponderosa Pine ecotype at removing and husking seeds from Lodgepole Pine cones, and vice versa. Divergent selection should be less evident when foraging on 'neutral' species such as Engelmann Spruce, because both pine ecotypes should have similar intake rates on its easily accessible seeds (Fig. 8.1).

Over six years, Cody and several field assistants gathered data on the two pine ecotypes breeding six times in mixed pine and four times in Engelmann Spruce. Three of the breeding episodes in mixed pine and three in spruce were during late summer to early fall (hereafter late summer), whereas three of the breeding episodes in mixed pine and one in spruce were during late winter to early spring (hereafter late winter). Breeding pairs were recorded following the methods used by Julie Smith in the South Hills and intake rates of each ecotype were measured opportunistically.

To quantify feeding tradeoffs, we used the difference in the mean intake rates between the two pine ecotypes foraging on the same conifer. This difference is illustrated as the difference between A and B on the left panel of Figure 8.2. We then standardized this difference by dividing it by the mean intake rate of the more efficient ecotype. This quotient provides a measure of proportionately how much less efficient one ecotype is relative to the other. For example, if this quotient or standardized difference were 0.3, it would mean that the less efficient ecotype removed and consumed seeds from cones 30% more slowly than its competitor. We termed this value as the performance differential. We reasoned that the larger the performance differential, the less likely the less efficient ecotype was to occur and breed near or with the more efficient ecotype.

Cody then examined whether and how different measures of reproductive isolation were related to the performance differentials. The performance differentials were small when foraging on spruce (0.02 and 0.07 in late summer and late winter, respectively, in Figure 8.3), but much larger when foraging on pine (0.30 and 0.51 in late winter and late summer, respectively, in Figure 8.3). The high values for late summer when foraging on pine (0.51) were expected because the crossbills were then foraging on closed, tough and woody-scaled cones of pine, which are the most challenging and most likely to accentuate the differences in performance between the two ecotypes (see Chapters 1 and 2). The average

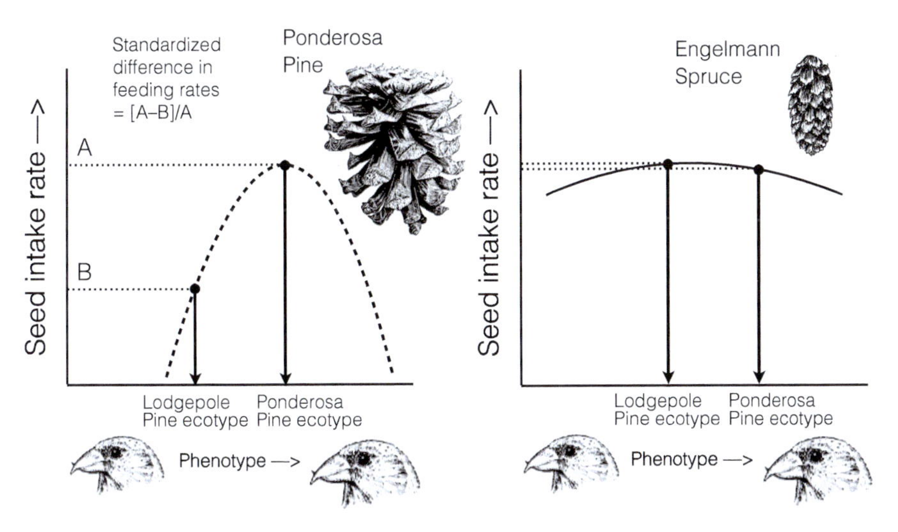

FIGURE 8.2 We estimated the relative seed intake rates for the two pine ecotypes by subtracting the mean intake rate of the less efficient ecotype (B; Lodgepole Pine ecotype foraging on Ponderosa Pine *Pinus ponderosa* in the left panel) from that of the more efficient ecotype (A; Ponderosa Pine ecotype foraging on Ponderosa Pine in the left panel), and then dividing by the latter (A). We refer to this quotient or standardized difference as the performance differential. We do not show the seed intake rate curve for Lodgepole Pine *P. contorta* to aid in visualization, and because usually only one of the two ecotypes foraged on their nonpreferred pine at any given location in one season. Modified from Porter and Benkman (2022).

performance differential for the four breeding episodes when foraging on spruce was 0.015 or approximately zero, indicating that the two ecotypes had nearly indistinguishable seed intake rates.

As expected, when the two ecotypes had similar intake rates (foraging on spruce), habitat isolation and the differences in the propensity of individuals of the two ecotypes to breed (we refer to the latter as 'reduced immigrant fecundity' in Figure 8.3) contributed little if anything to reproductive isolation. However, these two components of reproductive isolation increased markedly when the ecotypes fed on pine and were especially high for habitat isolation (habitat isolation averaged 0.83 for the six episodes when feeding on pine). Most striking are the high levels of behavioral isolation, such that behavioral isolation averaged only slightly less than total reproductive isolation, where total reproductive isolation is a measure of the sequential effect of all three isolating barriers (Ramsey et al. 2003; Smith and Benkman 2007). When the performance differential was greatest (0.51; pines in late summer), both behavioral isolation and total reproductive isolation were complete in two of the years (no mixed pairs) and almost complete in the other episode (Fig. 8.3; both behavioral and total reproductive isolation = 0.999; Porter and Benkman 2022). It is worth noting that Cody focused on locations where both ecotypes were common (Engelmann Spruce and mixed pine) and thus had

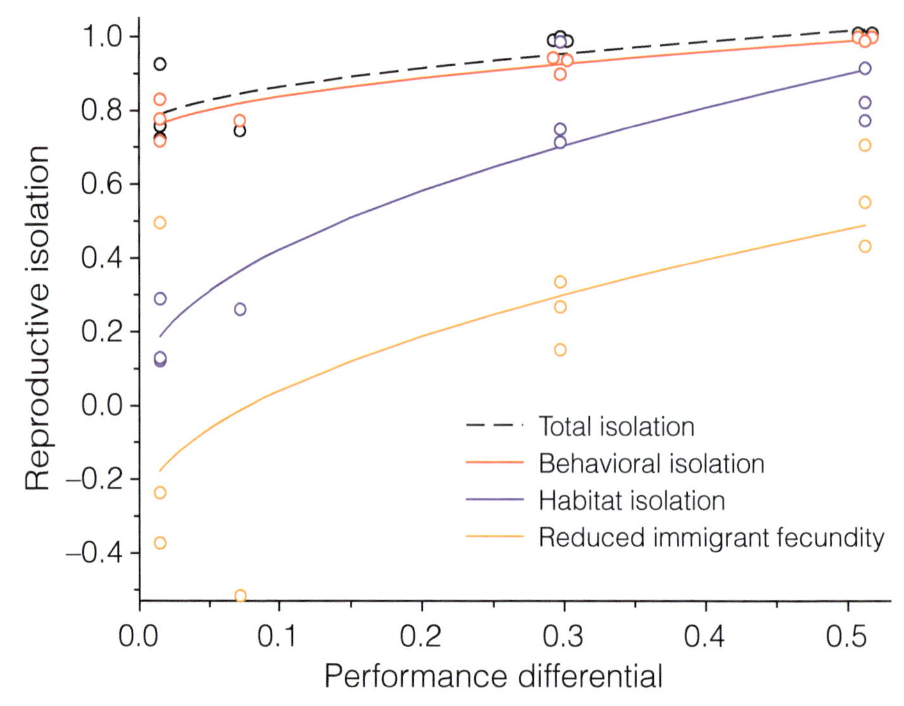

FIGURE 8.3 Total and different components of premating reproductive isolation between the pine ecotypes increase with increases in the performance differential (the standardized difference in seed intake rates, as in Figure 8.2), as predicted by ecological speciation. A value of zero for reproductive isolation = random mating; one = complete reproductive isolation; negative values imply a greater occurrence or potential for interbreeding than would occur by chance. The curves represent curvilinear regressions for total reproductive isolation (black line and circles; r^2 = 0.69, P = 0.003), behavioral isolation (red dashed line and circles; r^2 = 0.90, P <0.0001), habitat isolation (purple line and circles; r^2 = 0.86, P <0.0001) and reduced immigrant fecundity (orange line and circles; r^2 = 0.48, P = 0.03). Reduced immigrant fecundity is the term we use for the extent to which individuals of the less efficient ecotype breed (form mating pairs) relative to those of the more efficient ecotype. Data from Porter and Benkman (2022).

less potential for habitat isolation compared to vast expanses of either Lodgepole Pine or Ponderosa Pine where only one of the ecotypes generally dominates. We suspect that reproductive isolation would be highest in vast monospecific forests, such as those of Ponderosa Pine in northern Arizona and New Mexico. However, even in mixed pine, reproductive isolation was nearly complete.

Although we were unsuccessful in acquiring National Science Foundation support for this research, I found one comment by a reviewer especially useful. This reviewer noted that we did not have an alternative hypothesis to that of the performance differential influencing reproductive isolation. One such alternative hypothesis would be that absolute levels of seed availability (as measured by intake rates) affect reproductive isolation. During large flushes of resources, for example, the relative importance of feeding tradeoffs should be reduced because intake rates are elevated for all bill sizes, allowing a wider range of bill sizes to breed successfully. Conversely, when seed availability is low, we expect that the less efficient ecotype should be even less able to breed in the company of the more efficient ecotype. The basic idea is that increased productivity relaxes feeding tradeoffs and thereby the degree to which a given tradeoff impedes the use of alternative resources (Poisot et al. 2011). Cody estimated the necessary intake rates for a male to feed itself, its mate and their nestlings using the methods described in Chapter 2, 'Timing of breeding'. He then compared these values to mean intake rates for each ecotype under conditions of relative scarcity and abundance, respectively. He found exactly what we anticipated: The relationship between reproductive isolation and the performance differential was positive and strong only when relative seed availability was low (Porter and Benkman 2022). Increased productivity appears to relax feeding tradeoffs, thereby reducing reproductive isolation.

Overall, strong feeding tradeoffs, especially when seeds are not too plentiful, are critical for complete or nearly complete reproductive isolation. This is exactly the situation for the Cassia Crossbill in the South Hills, year after year. Conversely, the convergence of multiple ecotypes to nest while feeding on abundant, easily accessible seeds like those in Engelmann Spruce cones leads to interbreeding and potentially substantial gene flow. In addition, behavioral isolation is the strongest barrier and variation in behavioral isolation can largely account for variation in reproductive isolation (Fig. 8.3). I will address these points below, beginning with behavioral isolation and why it is related to the performance differential.

Assortative flocking and behavioral isolation between ecotypes

The rather high baseline levels of behavioral isolation (~0.75) even when performance differentials approximate zero (Fig. 8.3) presumably reflect mate preferences based on vocalizations and the tendency to flock assortatively. However, behavioral isolation increases to especially high levels as performance differentials increase. This increase in behavioral isolation is unlikely to reflect changes in mate preferences per se—at least, we haven't been able to envision a plausible reason for mate preferences to vary in such a manner—but could reflect an increase in assortative flocking. As discussed in Chapter 6, individuals benefit most from the foraging information provided by flockmates when all flock members have similar intake rates. When performance differentials are large, as when foraging on Ponderosa Pine (Fig. 8.2, left panel), then the two ecotypes have quite different average intake rates and individuals should avoid flocking with members of the other ecotype. However, when performance differentials approximate zero, as when foraging on Engelmann Spruce (Fig. 8.2, right panel), then individuals of both ecotypes have similar intake rates with little disadvantage resulting from flocking with members of another ecotype.

To test whether assortative flocking increases with increases in the performance differential, Cody recorded the occurrence of assortative and mixed flocks when searching for pairs. Figure 8.4A shows that indeed the proportion of flocks that were assortative increased with increases in the performance differential. Cody further tested whether this pattern was the result of an active and variable preference for association by using playbacks of contact calls during three breeding episodes. One of the breeding episodes occurred when crossbills fed on spruce in late summer and two when feeding on pine, with one in late summer and one in late winter. Cody found that both ecotypes decreased their propensity to land in response to heterotypic playbacks (those of the other ecotype) as compared to their propensity to land in response to playbacks of their own ecotype as the performance differential increased (Fig. 8.4B). This decline in the propensity to land in response to heterotypic playbacks provides a behavioral mechanism for the increase in assortative flocking as performance differentials increased (Fig. 8.4A). The increase in assortative flocking in turn can account for the increase in behavioral isolation as performance differentials increase (Fig. 8.4C). More assortative flocks lead to stronger behavioral (and total) reproductive isolation.

These results nicely spoke to our original motivation for studying the resource assessment benefits of flocking: its potential relevance to reproductive isolation (Chapter 6, 'Public information and assortative flocking'). It was now easy to envision that as divergent selection for foraging on alternative key conifers caused ecotypes to evolve different bill sizes and foraging abilities, natural selection would have further favored individuals that flocked assortatively by ecotype, especially when foraging on their key conifers. This in turn would have contributed to behavioral isolation that would lead to greater reproductive isolation beyond that arising from habitat isolation and reduced immigrant fecundity. The advantages of assortative flocking and rapid and accurate recognition of ecotypes would also have favored divergence in contact calls, as we have observed for the Cassia

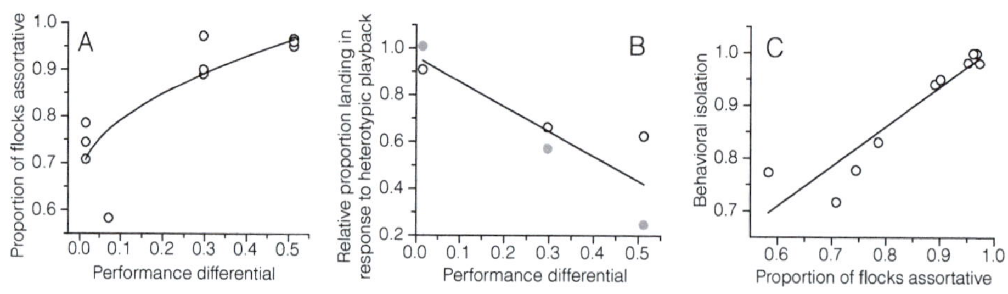

FIGURE 8.4 (A) The proportion of flocks that were assortative by ecotype increased as performance differentials increased ($r^2 = 0.69$, $P = 0.003$), presumably because (B) both Lodgepole Pine ecotypes (filled circles) and Ponderosa Pine ecotypes (open circles) were less likely to land to playbacks of each other's calls as performance differentials increased ($r^2 = 0.77$, $P = 0.022$). As expected, if behavioral isolation increases because of assortative flocking, (C) behavioral isolation increased with the proportion of flocks that were assortative ($r^2 = 0.79$, $P = 0.0006$). Modified from Porter and Benkman (2022).

Crossbill (Chapter 6, 'How might new ecotypes evolve?'). As contact calls diverge further, assortative flocking is likely to increase (Fig. 6.7; Porter and Benkman 2019), leading to increased reproductive isolation between ecotypes (Fig. 8.4C).

Our results also reveal that our previous finding of a decreasing propensity to land in response to calls of heterotypics as the difference in bill depth increases (Fig. 6.3B) depends on feeding on conifers with large performance differentials. Fortuitously, all the data included in Figure 6.3B were gathered when crossbills experienced large performance differentials. If we had instead measured crossbill responses to playbacks when the different ecotypes were feeding on (for example) Engelmann Spruce, we presumably would not have detected such a strong, graded response. Like much of our work, these results illustrate the value of being able to measure intake rates for both interpreting data and understanding crossbills more broadly.

Are consistently high levels of reproductive isolation unique to Cassia Crossbills?

Although reproductive isolation between Lodgepole Pine and Ponderosa Pine ecotypes was nearly complete when they foraged on pine (6 mixed pairs or 1.2% of 490 breeding pairs) and similar to that recorded for Cassia Crossbills (12 mixed pairs or 1.4% of 846 pairs involving Cassia Crossbills), interbreeding was an order of magnitude greater when the pine ecotypes bred while foraging on Engelmann Spruce (17 mixed pairs or 13.5% of 126 pairs). Given that some hybrids survive and mate with pine ecotypes, the resulting gene flow presumably accounts for why the pine ecotypes are not as genetically distinctive as the Cassia Crossbill (Parchman et al. 2016). The reproductive consequences of regular use of easily accessible seeds for breeding by most ecotypes likely also accounts for the limited genetic distinctiveness for the other ecotypes of Red Crossbills (Parchman et al. 2016).

We lack genetic data for the Newfoundland Crossbill *Loxia c. percna* and the Central American Crossbill *L. c. mesamericana*. There is reason to suspect both taxa are distinctive. For starters, the Newfoundland Crossbill is geographically isolated from areas with many Red Crossbills, potentially reducing opportunities for gene flow between ecotypes. However, the large, periodic seed crops of White Spruce *Picea glauca* on Newfoundland could lead to episodes of interbreeding if other Red Crossbills move onto Newfoundland during these large seed crops (its seeds appear to be equally accessible as those of Engelmann Spruce; see Fig. 4.9). Regardless, the Newfoundland Crossbill undoubtedly co-occurs with other Red Crossbills much less often than do (say) Lodgepole Pine and Ponderosa Pine ecotypes in the Rocky Mountains. It is thus reasonable to expect it is somewhat divergent relative to its congeners.

Of all the New World crossbills, the Central American subspecies seems the most comparable to the Cassia Crossbill, and perhaps for the same reasons it will eventually be found to be quite distinct genetically. In the highlands of Central

America where this crossbill mostly occurs, it apparently relies on seeds in the serotinous cones of *Pinus oocarpa* (J. van Dort, pers. comm. 2013; Benkman 2017b; Fig. 8.5A). Importantly, there is no conifer in its range with easily accessible seeds like Engelmann Spruce. This crossbill overlaps with Mexican Crossbills *L. c. stricklandi* in the northern part of its range but can be distinguished by vocalizations and by a combination of bill depth and wing length (Fig. 8.6A), as the former subspecies is larger on average in both dimensions. Notably, the wings are also disproportionately shorter in the Central American Crossbill, indicating it is more sedentary, like the Cassia Crossbill.

Intriguingly, variation in the mandible lengths of the Central American Crossbill (Fig. 8.6B) is similar in pattern to the seasonal variation in seed intake rates for the Cassia Crossbill (Fig. 7.3). This suggests that the Central American Crossbill experiences comparable seasonal variation in seed intake rates. Mandibles incur less wear and increase in length when seeds are more accessible. Conversely, mandible lengths decrease because of increased wear when seeds are more difficult to access. Given that both the Central American and Cassia Crossbill feed extensively on seeds in serotinous cones (Fig. 8.5), I suspect that the seasonal dynamics in *P. oocarpa* seed profitability is driven at least in part by seed depletion by crossbills. If so, then the similarities between these two crossbills could extend to high levels of reproductive isolation.

As an aside, the variation in mandible wear (Fig. 8.6B) might account for some of the variation in mandible lengths found by Howell (1972) among crossbills collected in Nicaragua. This variation apparently caused Howell to suggest that there were two bill size classes of Red Crossbills in Nicaragua. However, the variation in bill depth among the crossbills that have been collected in Central America is unimodal (Fig. 8.6A). Moreover, the range of variation is similar to that

FIGURE 8.5 Two crossbill taxa that rely on seeds in old weathered serotinous pine cones. (A) A Central American Red Crossbill *Loxia c. mesamericana* foraging on *Pinus oocarpa* cones in Honduras. (B) A Cassia Crossbill *L. sinesciuris* foraging on Rocky Mountain Lodgepole Pine *P. contorta* var. *latifolia* cones in the South Hills, Idaho. Photographs taken by John van Dort on 18 January 2014, and Ryan P. O'Donnell on 19 November 2014, respectively.

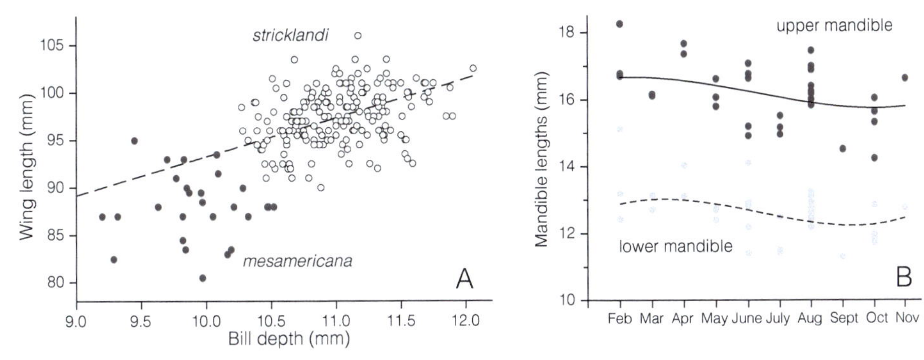

FIGURE 8.6 (A) The combination of wing length and bill depth distinguish Central American Crossbills *L. c. mesamericana* (n = 27) from Mexican Crossbills *L. c. stricklandi* (n = 193). Because these specimens lacked recordings of their vocalizations, I categorized individuals into their respective subspecies based on their measurements and locations of collection. Central American Crossbills have relatively short wings given that they lie below the dashed line, which represents the best-fit regression to data for nomadic North America ecotypes (n = 441) including the Mexican Crossbill but excluding the Central American Crossbill and Newfoundland Crossbill *L. c. percna*. (B) The lengths of the upper and lower mandibles of 32 Central American Crossbills vary seasonally over time. The curves represent polynomial equations fit to the data.

found in other call types: For example, the coefficient of variation for bill depth of male Central American Crossbills is 3.06, which falls within the range of the same measures for six North American call types (2.35–3.70; Groth 1993a). Together, these data indicate the occurrence of just a single ecotype in Central America besides, perhaps, occasional visiting Mexican Crossbills.

Although future studies are needed to determine to what extent our results can be extended to other call types of Red Crossbills, I would like to end this chapter by mentioning some examples from Europe and North Africa. First, we conducted genetic analyses on a subset of Old World crossbills (Parchman et al. 2018) similar to those we conducted on North American crossbills (Parchman et al. 2016). Our most striking result from this work was the finding that those crossbills that rely exclusively on seeds in a pine with serotinous cones (Aleppo Pine *Pinus halepensis*) were genetically distinct from other Red Crossbills, much like the Cassia Crossbill and its congeners. Two of these forms are well-recognized subspecies—*Loxia c. poliogyna* and *L. c. balearica*—that are also well isolated geographically, occurring in the Atlas Mountains of North Africa and on the island of Mallorca, respectively. Such geographic isolation weakens the case that their reliance on a stable food resource alone causes their genetic distinctiveness. However, a third, less widely recognized population in southern Spain (putatively *L. c. hispana*), is likely also reliant on Aleppo Pine and appears to be genetically distinct, although not quite as distinct as its more isolated relatives. No other European crossbill that we analyzed was particularly divergent (Parchman et al. 2018).

One puzzling result from genetic studies on crossbills in Europe are the few if any genetic differences between the largest of all crossbills, the Parrot Crossbill *L. pytyopsittacus*, and the substantially smaller Red (Common) Crossbill (Piertney et al. 2001; Parchman et al. 2018), whose distributions overlap widely in northern Europe (Newton 1972). Detailed studies in Scotland of these two taxa plus the intermediate-sized Scottish Crossbill *L. scotia* (Summers et al. 2007) found no interbreeding between Parrot and Red Crossbills, but did find both to occasionally hybridize with Scottish Crossbills (2 mixed pairs [4%] of 52 pairs based on contact calls; 5 mixed pairs [5%] of 93 pairs based on excitement calls; Summers et al. 2007). The Scottish Crossbill could thus act as a genetic bridge between Parrots and Reds. However, as the vast majority of Parrots and Reds (>99%) will never encounter a Scottish Crossbill, this appears unlikely to account for their overall genetic similarity. Perhaps Parrot Crossbills have diverged from other *Loxia* only recently, which combined with relatively large population sizes might serve to limit genetic divergence via genetic drift. Though plausible, I also wonder whether larger sample sizes of mated pairs would reveal more significant evidence of hybridization. There is one reason to suspect this might be the case. Both species regularly nest in spring when Scots Pine *Pinus sylvestris* and Norway Spruce *Picea abies* cones open. The seeds are then readily available, and the large performance differentials on closed cones are likely to diminish rapidly, potentially weakening reproductive isolating barriers. On the other hand, frequent hybridization is unlikely given the absence of intermediate bill sizes between Parrot and Red Crossbills measured from Fennoscandia (based on 249 birds; Summers et al. 2002). At this point, I can only hope that others take up the challenge, initiating a study like the ones Ron Summers orchestrated in Scotland and Julie Smith and Cody Porter undertook in North America, to further our understanding of the extent and conditions leading to assortative mating between crossbills in Europe.

Chapter 9

The Future of Crossbills: Climate Change and Other Threats

> One of the penalties of an ecological education is that one lives alone in a world of wounds. Much of the damage inflicted on land is quite invisible to laymen.
>
> (A. Leopold 1987: 286)

As much as I love observing and thinking about the natural history and diversity of crossbills, I find myself increasingly worried about their future. Like all too many other species, crossbills are experiencing numerous and oftentimes accelerating threats. One threat that I touched on in Chapter 5 was introduced species: in particular, the introduction of American Red Squirrels *Tamiasciurus hudsonicus*. As described earlier, Red Squirrels are widespread in the conifer forests of North America and efficient preemptive competitors that harvest vast quantities of closed cones, thereby reducing the availability of cone-seed to crossbills. The result is fewer crossbills and weaker interactions between crossbills and conifers. However, Red Squirrels are not everywhere. They are absent from areas where forests have become established while surrounded by barriers impenetrable to the species. This includes the island of Newfoundland where spruce forests expanded 9,000 years ago after the retreat of glaciers (Jackson et al. 1997), and the Cypress Hills where Lodgepole Pine *Pinus contorta* forests expanded around 4,600 years ago (Sauchyn and Sauchyn 1991; Sauchyn 1997) while isolated by a vast sea of prairie (Barnosky 1989). In the absence of Red Squirrels, crossbills reach high population densities and coevolve with conifers. Simultaneously, the conifers lose squirrel-directed defenses because of relaxation of selection by Red Squirrels. This loss of squirrel defenses in turn makes cone-seed quite vulnerable to their introduction. In Newfoundland, introduced Red Squirrels harvest much higher proportions of the cones than on the mainland and have most likely caused the massive decline of the Newfoundland Crossbill *Loxia c. percna* (Chapter 5). The same phenomenon has apparently occurred in the Cypress Hills (Chapter 5). Clearly, it is imprudent to introduce tree squirrels to areas with endemic crossbills.

One of the most notorious introduced species causing havoc for North American conifers is a fungal pathogen called White Pine Blister Rust *Cronartium ribicola*. White Pine Blister Rust has caused severe declines in some of the white or 'soft' pines (subgenus *Strobus*)—especially Whitebark Pine *Pinus albicaulis* and Limber Pine *P. flexilis* in western North America (Schoettle and Sniezko 2007). As a result, Whitebark Pine is now listed as a threatened species in the United States. Seeds of pines susceptible to Blister Rust are not 'key conifers' for crossbills in North America. Nonetheless, declines in Eastern White Pine *P. strobus*, whose seeds are commonly fed upon by Red Crossbills *Loxia curvirostra* in fall and winter (Benkman 1987a), would likely have adverse effects on the crossbill population. Introduced insects are also causing conifer declines and reductions in cone-seed. For example, introduced Seed Bugs *Leptoglossus occidentalis* in the French Alps are damaging up to 70% of the seeds in some high-elevation stands of two important pines for crossbills, Scots Pine *P. sylvestris* and European Black Pine *P. nigra* (Lesieur et al. 2014). In the eastern United States, non-native Hemlock Woolly Adelgids *Adelges tsugae* and the Elongate Hemlock Scale *Fiorinia externa* are killing hemlocks (*Tsuga canadensis* and *T. caroliniana*) (Toenies et al. 2018).

For resident populations of crossbills, loss of mature conifers on which crossbills rely is likely to have a proportionate effect. That is, a 20% loss of mature forest is likely to cause a comparable decline in the crossbill population. For nomadic populations, however, forest loss has the potential to have a disproportionate impact (Benkman 1993b). In some years nomadic populations might be limited to a small fraction of their range-wide distribution (Chapter 2). If forest loss is concentrated in that fraction, then crossbills could incur declines greater than the overall proportionate loss of forest. Thus, a 10% loss of total forest could cause a 50% decrease in the crossbill population in a given year when the forest loss represents 50% of all the forest that could have a good cone crop that year. Crossbills might rebound quickly in subsequent years when large seed crops are produced in less impacted regions, but non-uniform forest loss has the potential to cause lower population lows along with greater annual fluctuations in population size, which could be problematic.

Widespread harvest of the conifers they rely on also threatens crossbill populations. I first became concerned with logging that resulted in concentrated regional forest loss in 1988. At that time, I was studying White-winged Crossbills *Loxia l. leucoptera* foraging on a large Red Spruce *Picea rubens* seed crop in central New Brunswick, Canada, where a network of roads was being graded in a grid to provide extensive and ready access for loggers in both the near-term and for decades to come. I have not returned to the area since the spring of 1989. However, after the extensive clearcuts I observed that year, this area is unlikely to support many crossbills as long as the interval between successive clearcuts—i.e., the harvest rotation—is sufficiently short, such that the trees are unlikely to produce large seed crops before they are cut (Fig. 9.1). Generally, few cones are produced during the first several decades of a conifer's life. After that, seed crops increase in size and regularity with tree age and size, with some species not producing their largest seed crops until they are one or more centuries old (Burns and Honkala 1990).

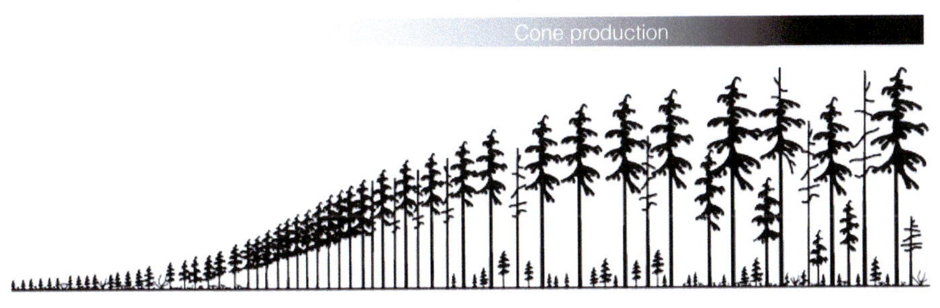

FIGURE 9.1 The size and frequency of seed crops increase with conifer size and age, as indicated by the horizontal bar where darker shades represent larger and more frequent seed crops. If forests are clearcut (or experience a stand-replacing fire) midway through the time interval depicted, then few if any large seed crops would occur. In contrast, if disturbances do not occur until the end of the depicted time interval, then large seed crops would be produced during a larger fraction of time. Modified from Kimmins (2003) with permission from *The Forestry Chronicle*.

Thus, the shorter the rotation interval, the greater the proportion of time a given location will be unproductive for crossbills (Fig. 9.1; Benkman 1993b). Moreover, Balsam Fir *Abies balsamea* dominates many younger stands in New Brunswick (Betts et al. 2022), a habitat that is effectively a desert for white-wings regardless of the forest's age because they only occasionally eat fir seeds (Chapter 2, 'Timing of breeding'). Although the effect of these forest policies on the long-term population trends of white-wings may be limited if relatively little Black Spruce *Picea mariana* is affected, such management has prevented forests from acquiring old-growth characteristics over large swaths of New Brunswick and elsewhere in the Acadian region, causing substantial declines in over 50 species of breeding birds between 1985 and 2020 (Betts et al. 2022). Unfortunately, extensive clear-cutting is not limited to New Brunswick and has occurred across Canada including in Ontario and Quebec, likely with dire consequences for a diversity of native species (Mackey et al. 2023).

Climate change, fire and beetle kill

The most insidious threat facing crossbills is climate change, as it is causing an accelerating decline in both the quality (suitability) and area of mature and old-growth conifer in both protected and unprotected areas. The decline in mature forest suitability is the subject of the two subsequent sections in this chapter. Here I focus on mature forest declines.

One of the starkest and for humans frightening effects of climate change is the increase in the size, frequency and severity of fires (Westerling et al. 2006; Barbero et al. 2015; Abatzoglou and Williams 2016; Parks and Abatzoglou 2020). The immediate impact of high-severity, stand-replacing fire on crossbills is the loss of cone-producing forests. As in the aftermath of a clearcut, the fewer trees producing cones, the less food for crossbills; less food for crossbills means *fewer* crossbills. Frequent high-severity fires act to prevent trees from reaching cone-producing age, or at least reduce the proportion of time that forests are of cone-producing age, with inimical effects on crossbills much like those from short-harvest rotations (Fig. 9.1).

Conifers such as Lodgepole Pine that produce serotinous cones are adapted to stand-replacing, high-severity fires and generally begin forming cones at younger ages compared to other conifers (Burns and Honkala 1990). Thus, we might expect serotinous species to not suffer as much from increasing frequencies of high-severity fire (Davis et al. 2023). However, fires over the last few decades have begun occurring frequently enough that they often reburn a given habitat patch before Lodgepole Pine develops a sizeable canopy seed bank, which requires 40–70 years (Schoennagel et al. 2003). When this occurs, it substantially depresses tree regeneration (Turner et al. 2019). The decrease in the canopy seed bank because of reoccurring disturbances such as fire has been referred to as the immaturity risk or interval squeeze (Keeley et al. 1999; Johnstone and Chapin 2006; Brown and Johnstone 2012; Enright et al. 2015) and contributes to the shift in forest dominance from conifers such as semi-serotinous Black Spruce to broad-leaved trees in the boreal forests of North America (Whitman et al. 2019; Baltzer et al. 2021). More frequent fires are also projected to cause a decline in the occurrence of Lodgepole Pine in the Greater Yellowstone region (Westerling et al. 2011; Turner et al. 2022). As we've discussed in Chapters 2, 4 and 5, Black Spruce and Lodgepole Pine are two 'key conifers'. Their decline will have adverse effects on their associated crossbill populations.

Warmer, drier conditions associated with climate change have also contributed to widespread increases in drought-induced tree mortality in the absence of fire (van Mantgem et al. 2009; Allen et al. 2010; Andrus et al. 2021), something predicted to increase in frequency, severity and duration throughout much of North America (Naumann et al. 2018). These conditions are also favorable for Mountain Pine Beetle *Dendroctonus ponderosae* (Bentz et al. 2010), which in the last several decades has killed immense numbers of mature Lodgepole Pine in the Rocky Mountain region (Fig. 9.2; Raffa et al. 2008; Audley et al. 2020; Hicke et al. 2020). With increasing temperatures and drought, Bark Beetles *Dendroctonus* and *Ips* spp. are killing increasing numbers of conifers in North America and Europe (Raffa et al. 2008; Allen et al. 2010; Jaime et al. 2024). The Southern Pine Beetle *D. frontalis* in the eastern United States is expanding northward as conditions farther north become more climatically favorable, with the potential to spread and kill vast numbers of pine (Lesk et al. 2017). Again, and quite simply, fewer mature cone-bearing conifers means fewer crossbills.

Not only is climate change directly and indirectly responsible for increases in mature conifer mortality, but increasing temperatures and drought associated with climate change are creating conditions that are adverse for forest regeneration

FIGURE 9.2 Lodgepole Pine *Pinus contorta* killed by Mountain Pine Beetles *Dendroctonus ponderosae* in the Rocky Mountains in northern Colorado. (A) Medicine Bow-Routt National Forest, 16 August 2009; (B) Rocky Mountain National Park, 3 July 2010.

(Davis et al. 2019, 2023; Coop et al. 2020; Stevens-Rumann et al. 2022). Conifer seedlings are especially vulnerable because they require a narrower set of conditions to survive than mature individuals (Bell et al. 2014). In addition, seed sources are fewer following the increasingly common large, high-severity fires, and thus little to no regeneration may occur after such a fire. This appears to be an escalating problem for Ponderosa Pine *Pinus ponderosa* and Douglas-fir *Pseudotsuga menziesii* (Davis et al. 2019, 2023; Stevens-Rumann et al. 2022). Even in the absence of fire, Ponderosa Pine is regenerating poorly if at all in many warmer sites (Petrie et al. 2023).

As described in Chapter 5, 'Lodgepole Pine cone-seed as a resource for crossbills and the effect of Red Squirrels', Lodgepole Pine seedling abundance after a stand-replacing fire is an index of crossbill carrying capacity in pre-fire mature stands, and thus a decline in seedling abundance likely reflects a decline in crossbill abundance. However, because seedling declines are also increasingly likely to reflect drought conditions during germination and early growth (Coop et al. 2020; Davis et al. 2023), seedling densities as an index will increasingly underestimate crossbill densities in pre-fire forests. Nevertheless, seedling decline—whether from drought or immaturity risk—is likely to adversely affect the number of crossbills that can be supported in the future. Indeed, the loss of mature cone-bearing conifers and the decline in their regeneration in many western North American forests (Allen et al. 2010; Coop et al. 2020; Stevens-Rumann et al. 2022; Davis et al. 2023) is a problem for crossbills that is likely to only worsen.

Cadieux and colleagues (2020) provided an estimate of the impact of climate change in combination with current and hypothetically more elevated levels of logging on population trends of songbirds in the boreal forest in northeastern Alberta, Canada. They used the relationships between the abundances of 72 songbird species and forest occurrence, composition and age to forecast bird abundances based on projected changes at both the stand and landscape level. The total area of conifer forest and its age were projected to decline substantially by the year 2100 and, not surprisingly, bird species relying on mature conifers were projected to decline along with it. Among the species included in their study was the White-winged Crossbill, which was projected to decline between 40% and 50%. As bad as this is, I strongly suspect that it overestimates the future carrying capacity for crossbills in these forests and thereby underestimates future declines, because it does not capture future declines in the *suitability* of remaining mature conifer forests for crossbills. Why this is the case is the focus of the next two sections.

Climate change, Cassia Crossbills and early seed shedding

In the first few years following my discovery of the Cassia Crossbill, I was not particularly worried about immediate threats to its persistence despite the restricted range of this species. Logging had been going on in the South Hills and Albions for decades but remained at a small scale. Fire did not seem like a great problem either: I was informed by a forester in the Sawtooth National Forest that individual blazes in the South Hills tended to be of limited size, because the pine forests were patchy within a matrix of sagebrush steppe (Fig. 5.4) that tended not to burn (Baker 2006). Indeed, the historical absence of massive fires that might have burned the entirety of pine forest in one or both mountain ranges presumably accounted for the long-term persistence and independent evolution of Cassia Crossbills (Chapter 5). Finally, American Red Squirrels had not been introduced here, and given the isolation of the area and our improved understanding of the impacts of invasive species, I didn't anticipate that they would be. If Red Squirrels *were* introduced, I also felt they could be eliminated reasonably quickly, due to the accessibility of most of the forests in the South Hills and Albions and the fact that their patchy nature would slow the spread of the invader.

Having concluded the Cassia Crossbill did not need pressing conservation attention, our initial work instead focused on basic ecological and evolutionary questions. For example, I was interested in documenting the dynamics of what seemed like a stable population to provide a contrast to the large year-to-year fluctuations found most elsewhere among crossbills (Chapter 2). To aid in this endeavor, Trevor Fetz and Julie Smith began conducting point counts as part of their dissertations in 2000 and continued fieldwork through 2002. In 2003, Lenny Santisteban took over the point counts surveyed by Trevor and Julie. These point counts had proven useful in providing a measure of habitat isolation between the different forms of crossbills in the South Hills (Chapter 7), and our goal was to continue measuring the extent of reproductive isolation between Cassia Crossbills and the two pine ecotypes. To our surprise and worry, we began noticing a decline in Cassia Crossbills after 2003 (Fig. 9.3). I was initially concerned that Mountain Pine Beetles—then devastating vast areas of Lodgepole Pine in the west including in mountains near my home in Laramie, Wyoming (Fig. 9.2)—were responsible for declines in Lodgepole Pine in the South Hills as well. However, this proved not

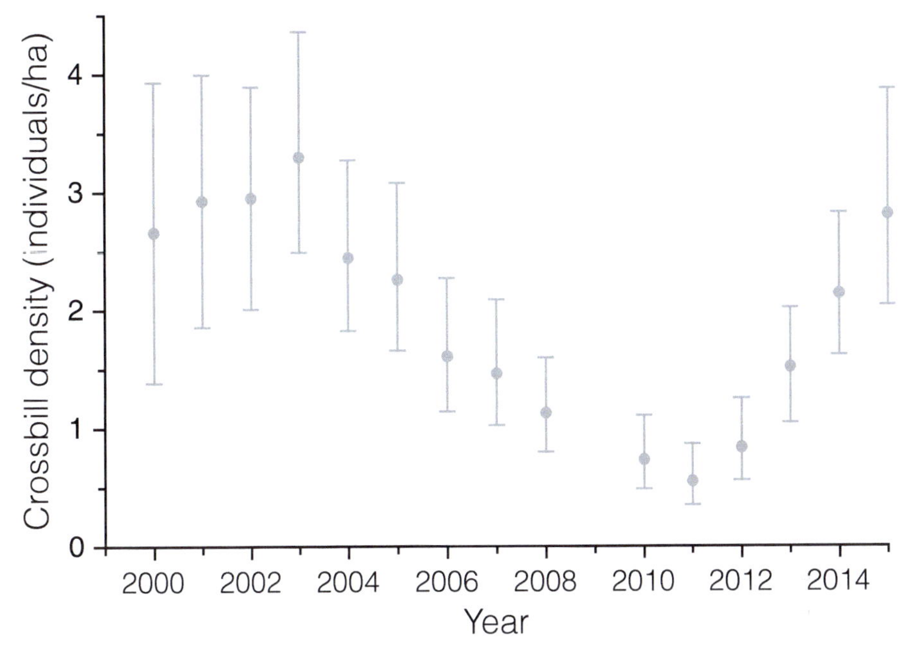

FIGURE 9.3 Cassia Crossbills began declining in abundance in the South Hills after 2003, then rebounded after 2011. Crossbill densities were based on the number of Cassia Crossbills observed perched during 10-minute point counts ($n = 68$ survey points in 2000, 74 survey points thereafter). Point counts between 2000 and 2002 were conducted in July and estimates were based on detections within a 50 m radius. In all subsequent years, point counts were conducted in late July to early August and estimates were based on distance-sampling methods. 95% confidence limits are plotted around the mean. Point counts were conducted in 2009, but the observers did not distinguish perched from flying birds; we include only perched crossbills in the estimates.

to be the case, and has yet to occur (Benkman 2016). If it wasn't elevated tree mortality, what was causing the crossbill decline?

To try to figure out what might have contributed to the population decline, Lenny combined point-count data (to estimate population density) with banding data (to estimate apparent annual survival). He found that the best predictor of the Cassia Crossbill's decline up to 2008 was mean spring temperature (Santisteban et al. 2012). The higher the spring temperature, the lower the annual survival; estimated declines in survival from this model could account for the observed decline in the population. Other factors—in particular West Nile Virus, which had become prevalent in the United States and had caused populations of other bird species to plummet—could be eliminated as a cause of crossbill decline (Santisteban et al. 2012). Though the results were compelling, I could not envision a mechanism by which higher spring temperatures adversely affect Cassia Crossbills. I initially wondered if higher spring temperatures could be related to increases in freezing rain, but couldn't find any evidence to support this supposition. I thought of freezing rain because following such an event at a field site in eastern Canada, White-winged Crossbills seemed to desert the area, presumably because cones covered with ice were impenetrable to the birds. Alternatively, higher temperatures might adversely affect crossbill breeding success, a pattern seen in some summer-breeding species. However adverse effects seem limited to those nesting in open habitats, with little effect on those nesting in forests like crossbills (Lauck et al. 2023).

We ultimately argued that the correlation between spring temperature and annual survival was spurious, and, instead, we focused on a less strongly supported model. Namely, we thought that lagged effects from hot summer days (≥32°C or 89.6°F) in the previous three to five years were responsible for the detected crossbill decline (Santisteban et al. 2012). This hypothesis fit with a sharp increase in the number of hot days in 2003 (Fig. 9.4), after which the crossbill population began plummeting (Fig. 9.3). Importantly, I could envision a plausible mechanism: Serotinous cones experiencing direct sunlight during hot summer days might open and shed their seeds prematurely, thereby reducing the abundance of cone-seed that would otherwise gradually become available to crossbills over the next several years. Hot, dry weather events have also been found to initiate seed release by serotinous cones of Aleppo Pine *Pinus halepensis* (Nathan et al. 1999). I reasoned

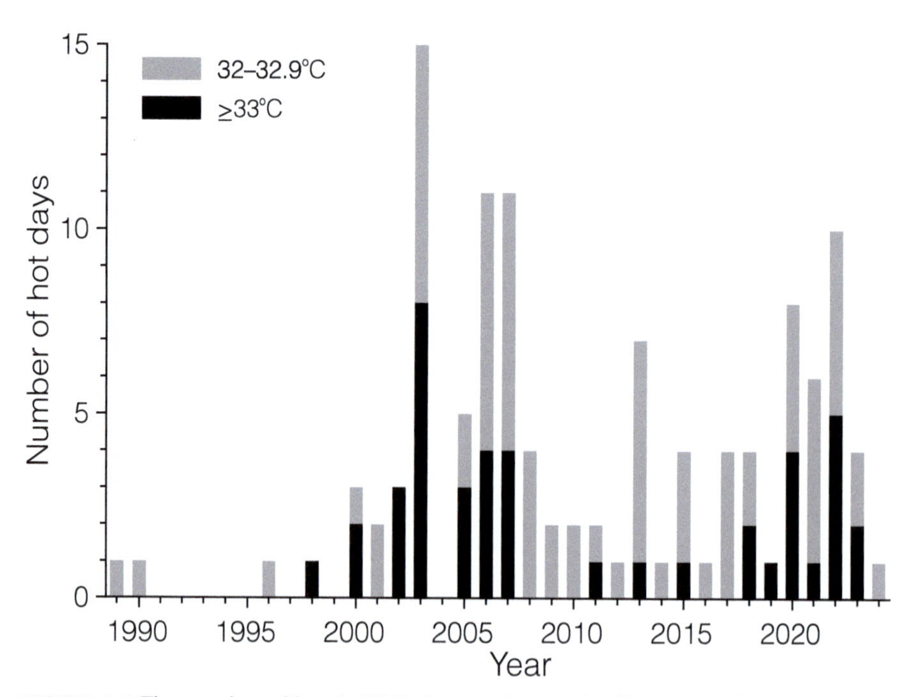

FIGURE 9.4 The number of hot (≥32°C) days in the South Hills, Idaho (data from Magic Mountain SNOTEL site, Station ID: 610) between 1989 and 2024. These data, accessed at https://www.nrcs.usda.gov/ on 5 February 2025, show an increase in temperatures and number of hot days compared to that depicted in Benkman (2016), for unknown reasons. Note the increase in daily maximum temperatures beginning in 2000.

that the cones most susceptible to opening from hot temperatures were the same ones that crossbills were most likely to rely on over the ensuing three to five years. That is, cones that had weathered or would soon weather with slight gaps forming between the scales were prone to open and rapidly shed their seeds.

After publishing our paper (Santisteban et al. 2012), an extended stretch of summers with relatively few hot days (between 2008 and 2019; Fig. 9.4) provided an opportunity to test our hypothesis (Benkman 2016). If crossbills had declined because of numerous hot summer days during the previous three to five years, then Cassia Crossbills should rebound in subsequent years. The assumption for this prediction is that in the absence of a series of hot days, the canopy seed bank will gradually recover, allowing for an increase in the crossbill population. Other students continued conducting point counts where Julie, Trevor and Lenny had worked. As expected, crossbills rebounded after 2011 (Fig. 9.3).

An analysis of the annual change in crossbill population size in relation to the number of hot days (≥33°C) in the previous four years indicates that crossbills decline when there are four or more hot days but increase when there are fewer than three hot days (Fig. 9.5). The relationships between population change and

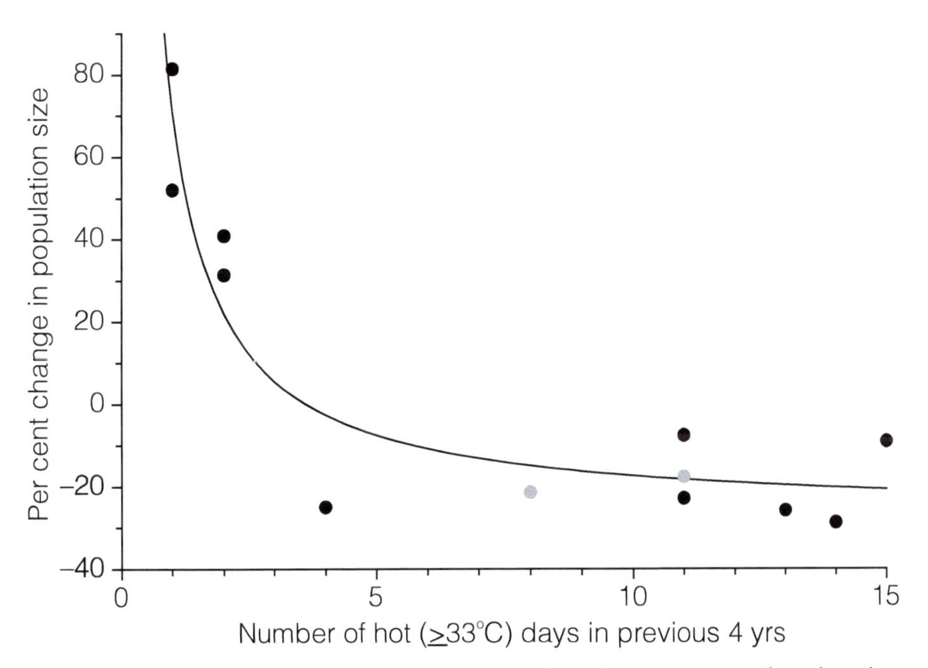

FIGURE 9.5 Crossbills decreased in population size when there were more than three hot days (≥32°C) in the previous four years, whereas they increased when there were less than three hot days. The curve represents the following regression: *% population change* = −27.19 + 98.37/ [*# hot days in previous 4 yrs* + 0.0001], $r^2 = 0.87$, df = 8, $P < 0.0001$). Gray symbols represent estimates of population change assuming that the population density in 2009, for which we don't have comparable population estimates, was intermediate between that in 2008 and 2010 (see Fig. 9.3); these two estimates (changes between 2008 and 2009, and 2009 and 2010) were not included in the regression.

the number of hot days in the previous three or five years are weaker (both r^2 <0.75) than the relationship between population change and the number of hot days for the previous four years ($r^2 = 0.87$). Similar patterns are found using 32°C as the temperature threshold. Spring temperature, which had been the best predictor of annual survival and population change in prior analyses, is uncorrelated with the annual change in population size. Changes in the size of annual seed crops, such as declining cone crops following high temperatures and increasing aridity, as found in Two-needle Pinyon *Pinus edulis* (Wion et al. 2020), could also be eliminated, as we did not detect a systematic change in annual seed crop size over our study (Fig. 7.2; Benkman 2016).

My interpretation of Figure 9.5 is that the cumulative effect of four or more hot days (≥33°C) in one summer (2003, 2006 and 2007; Fig. 9.4) caused seeds that would normally become accessible over the next four years to be shed prematurely. This could cause the approximately 20% annual decline in crossbills during the following four years and suggests that repeated exposure to temperatures around or especially above 33°C was the point at which populations were impacted. In 2003 for example, a three-day period had maximum daily temperatures of 33.4, 34 and 35°C, while in 2006 a three-day period had maximum daily temperatures of 36, 34.8 and 34.8°C. The fact that declines were only detected after multiple hot summer days perhaps reflects the importance of an increasing duration of high temperatures for cone opening when near the cone-opening temperature threshold (Alexander and Cruz 2012).

Although the results were suggestive (Fig. 9.5), these 'hot day' ambient temperatures were still well below the 45–60°C heat believed to be required to open serotinous Lodgepole Pine cones (Perry and Lotan 1977). However, another mechanism was possible: Arnit (1968 in Critchfield 1980: 12) suggested it was direct sunlight during hot days that caused cones within trees to open. Indeed, it was my own observations of trees in the South Hills whose closed, serotinous cones had opened following 2003 that made me think hot summer days played a role in the first place. At this point, no one had quantified whether cones within tree canopies experience cone-opening temperatures when air temperatures reach or exceed 32°C. The question was thus whether cones were likely to reach such temperatures in direct sunlight on hot days. There was reason to suspect this was the case: Cones on branches that have fallen to the ground and receive direct sunlight commonly open during warm summer days, attracting numerous seed-eating birds and mammals. But temperatures near the ground exceed those experienced in the canopy.

To evaluate whether cones within trees experience cone-opening temperatures, we first characterized the cone-opening temperatures for nearly 1,000 cones from 62 trees in the South Hills. My daughter Katherine measured the temperature at which cone scales began to open for the entire sample. This was convenient for her—I brought home a digital drying oven—although a tedious summer job. She set the oven to 40°C, and after maintaining that temperature for 20 minutes, she

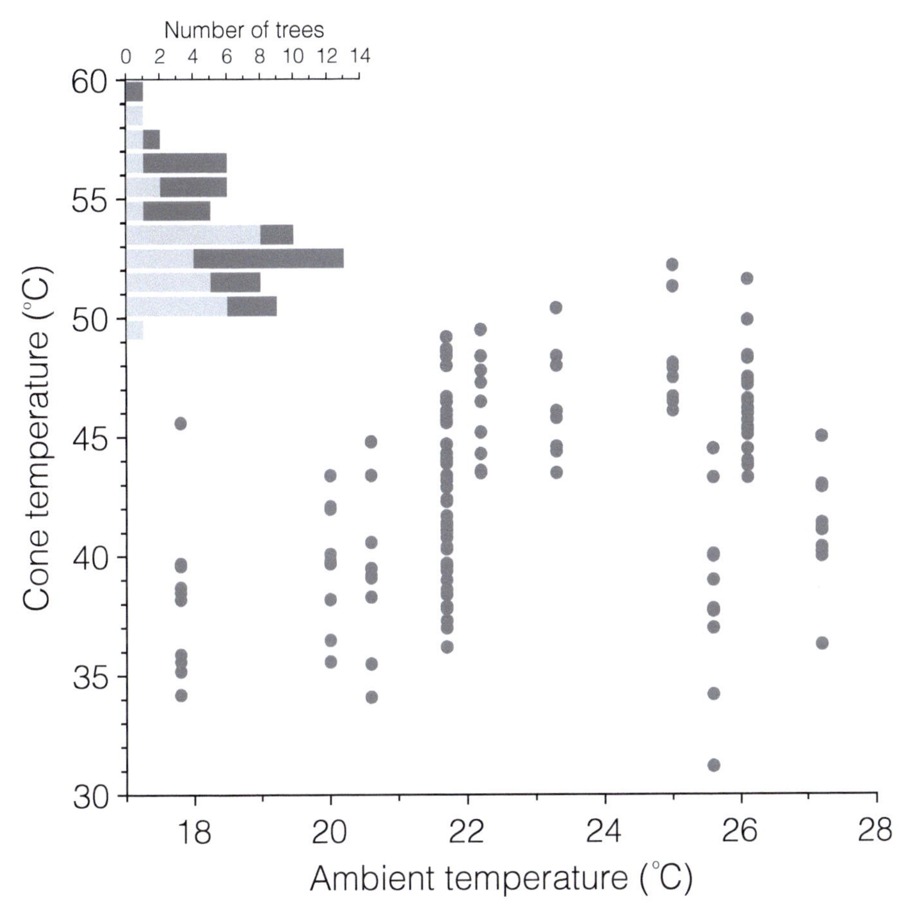

FIGURE 9.6 Maximum surface temperatures of cones (gray circles) in relation to air temperatures at the Magic Mountain SNOTEL site, June 2015. The histogram in the upper left shows mean cone-opening temperatures for 62 trees based on experiments using an oven: light gray bars represent values for 1–10-year-old cones for 30 trees; dark bars represent values for 15–20-year-old cones for 32 trees. Modified from Benkman (2016).

raised it 2°C and allowed it to stabilize for another 10 minutes. This was repeated for each 2°C increment in temperature until all the cones in the oven had opened or 60°C was reached, the maximum temperature for the oven. I then calculated the mean temperature at which a given tree's cones began to open as its cone-opening temperature. The distribution of cone-opening temperatures for the 62 trees is shown in the upper left of Figure 9.6.

To determine whether in direct sunlight the cone surfaces reach cone-opening temperatures, I recorded the maximum surface temperature of cones in direct sunlight using a FLIR T420 thermal-imaging camera (Fig. 9.7). I selected a cone-laden Lodgepole Pine growing in the open and measured the temperatures of cones approximately 1.5 to 2 meters above ground. I recorded the maximum

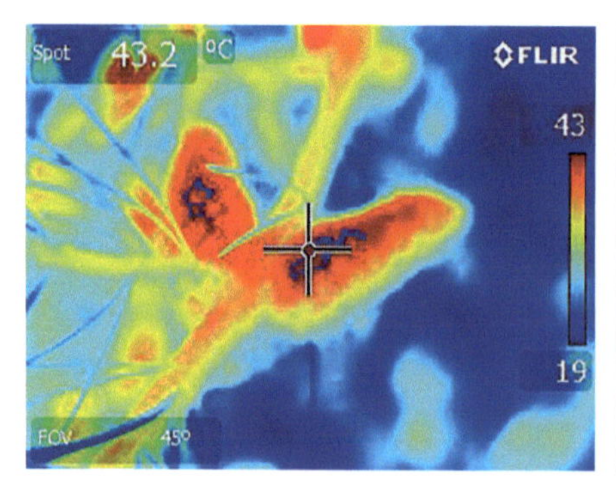

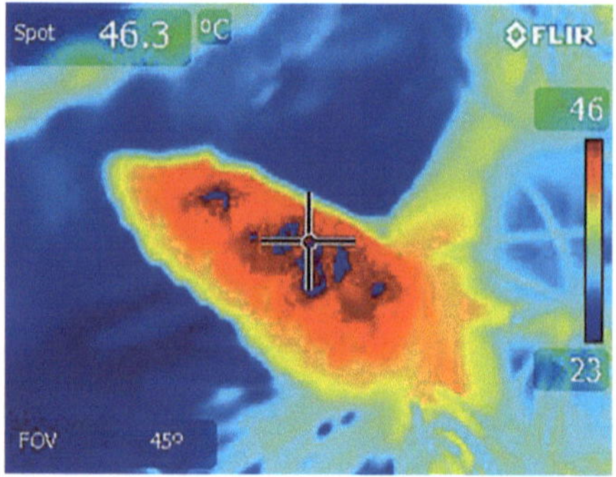

FIGURE 9.7 Thermal images of Rocky Mountain Lodgepole Pine *Pinus contorta* var. *latifolia* cones in direct sunlight in June 2015 at Porcupine Springs in the South Hills. The temperature in the middle of the cross is indicated in the upper left corner. From Benkman (2016).

temperature on the surface of 10 haphazardly chosen cones and compared these temperatures to the respective air temperature recorded every half-hour 1.6 miles (2.6 km) away at the Magic Mountain SNOTEL site. These data are plotted in Figure 9.6. Cone temperatures increased with increases in air temperature up to 25°C, then decreased (Fig. 9.6). I surmised that the decline at higher temperatures was probably the result of wind causing turbulence over the cones.

To test whether cone temperatures increase more continuously with increases in ambient temperature in the absence of wind, I measured cone temperatures on a nearly windless day when ambient temperatures reached over 34°C in my backyard after I moved to northern Colorado. I attached 10 closed cones to a dowel that was secured to a tripod around 1.5 meters above a lawn in an open area. Although the absolute cone temperatures do not necessarily reflect the temperatures cones would experience attached to a live branch in a canopy, a linear increase in cone temperature with increasing ambient temperature was detected (Fig. 9.8). I interpret this result as indicating that the progressive increase in cone temperatures up to 25°C in Figure 9.6 would continue at higher temperatures in the absence of wind.

I note that the maximum surface temperatures are limited to a small fraction of the cone surface (Fig. 9.7), and only some cones experience direct sunlight during the hottest times of the day. Nevertheless, the results (Figs 9.6 and 9.8) suggest the potential for an increasing overlap between cone temperatures and cone-opening temperatures as air temperatures increase. Such an overlap might be sufficient to cause cone scales to open when ambient temperatures exceed 33°C during multiple days. A recent study by Wyse and colleagues (2019) on Monterey Pine *Pinus radiata* provides another example of serotinous cones opening at high ambient temperatures; cone temperatures were around 15°C above the ambient temperatures, a similar difference to

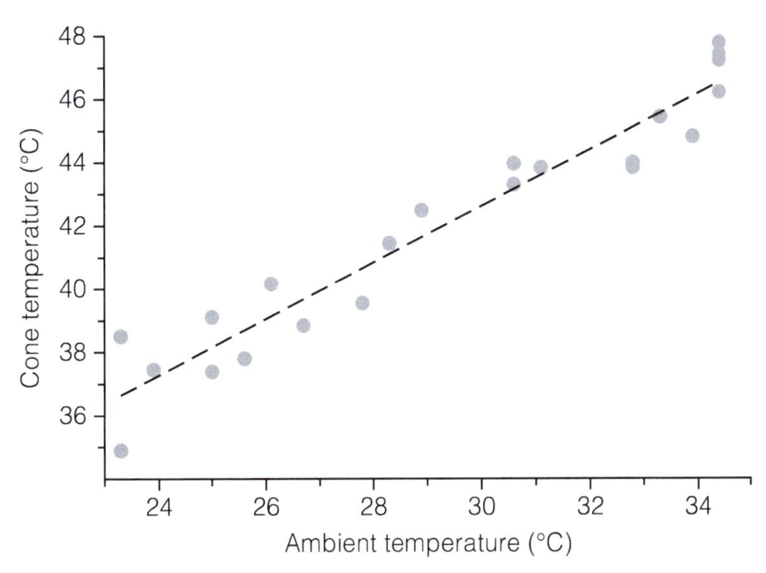

FIGURE 9.8 Mean maximum surface temperatures of 10 Rocky Mountain Lodgepole Pine *Pinus contorta* var. *latifolia* cones increased linearly with increasing ambient temperature when wind turbulence was limited. Cones were attached perpendicular to a wooden dowel around 1.5 m above a lawn. Air temperatures were from a weather station 370 feet (113 m) away during a calm day (wind speed <16 km/hour) on 11 July 2019. The dashed line is a best-fit linear regression ($r^2 = 0.93$, $P < 0.0001$). Wind speed had a negative effect on cone temperature in a generalized linear model, as expected; however, its effect was marginal ($P = 0.074$).

the one I observed for Lodgepole Pine (Figs 9.6 and 9.8). Cone opening would seem especially probable for weathered cones already prone to open, thereby depleting a fraction of the seeds likely to become accessible to crossbills over the next several years.

Although Cassia Crossbills had largely recovered by 2015 (Fig. 9.3), the South Hills experienced four hot summer days in 2020 and then five such days in 2022 (Fig. 9.4). This provided a further test of the hypothesis, albeit an unfortunate one for the crossbills. Compounding this heat, September 2020's Badger Fire (Fig. 5.7A) burned about 20% of all Lodgepole Pine in the South Hills and Albions, a poignant reminder of the occurrence of more frequent large, high-severity fires (Westerling et al. 2006; Barbero et al. 2015; Abatzoglou and Williams 2016). Assuming the 2020 Cassia Crossbill population was 3,607 individuals based on a revised estimate for 2016 from Behl and Benkman (2018), I projected a 20% decrease to 2,885 individuals as a direct result of the Badger Fire, followed by a further six years of annual declines of 20% due to hot summer days in 2020 and 2022. By 2023, then, there should have been only 1,477 crossbills remaining (that is, 2,885 × [0.8]3). This predicted number is just a bit greater than what colleagues have estimated for 2023. Encouragingly, the population may have rebounded somewhat in 2024, which is much earlier than I expected. This indicates that crossbill declines after a series of hot summer days are more nuanced than what the earlier analyses implied.

The outlook for the Cassia Crossbill is not promising given the increasing occurrence of high-severity fires and the increasing risk of hot summer days (Duffy and Tebaldi 2012; Christidis et al. 2015)—factors that will reduce the total area of mature forests by favoring plants other than Lodgepole Pine (Westerling et al. 2011; Turner et al. 2022) and diminish the carrying capacity of those that remain. The increase in fire danger between 1979 and 2020 has been especially great at higher elevations, particularly in the Basin and Range ecoregion of western North America (Alizadeh et al. 2023), the region where the Cassia Crossbill resides. In addition, increasing drought in the Cassia Crossbill's already relatively warm and xeric range is likely to cause failures in seedling recruitment and poor tree growth and survival (Hansen and Turner 2019; Turner et al. 2019). Indeed, Lodgepole Pine has been projected to disappear from the South Hills and Albions as well as from extensive areas to the north and east by the end of this century (Coops and Waring 2011). Although there are some opportunities for Lodgepole Pine's range to expand within the South Hills and Albions, locations with cooler microclimates and a suitable substrate for Lodgepole Pine growth are limited. Nonetheless, if we are successful at slowing and eventually reversing our current trajectory of increasing temperatures, I hold onto the hope that there are opportunities for the Cassia Crossbill to persist.

Higher temperatures also cause cones of non-serotinous conifers (e.g., Douglas-fir and Ponderosa Pine) to more rapidly shed seeds (Schopmeyer 1974). Thus, it seems likely that increasing temperatures will more widely reduce the availability of seeds in cones in late winter and spring, the period when seeds appear to be most limiting to crossbills (Chapters 2 and 4). Suggestive of increasing seed limitation, Red Crossbills in late winter and spring are coming to bird feeders (number of crossbills per day per feeder) in the Rocky Mountain region at an accelerating rate, with a greater than 20-fold increase between 1988 and 2010 (data from the Cornell Laboratory of Ornithology's Project FeederWatch). This increase of crossbill sightings at feeders occurred in spite of a concurrent decrease in their population as assessed along Breeding Bird Survey routes (Sauer et al. 2014). I have yet to examine more recent data. However, these findings may shed light on another crossbill mystery: what accounts for the restricted distribution of the Parrot Crossbill *Loxia pytyopsittacus*?

What limits the occurrence of Parrot Crossbills?

> The range of the Parrot Crossbill is surprisingly restricted ... considering that its main food-plant grows over more than six times the area occupied by the bird itself.
>
> (I. Newton 1972: 87)

Different crossbill taxa generally occur throughout the range of their key conifers, but sometimes beyond—*way* beyond. The occurrence of crossbills far beyond their usual range limits usually follows a combination of population increase and widespread cone failure (Koenig and Knops 2001), resulting in large numbers of individuals moving outside their typical distribution and (at times) into areas with few conifers. These large-scale and often somewhat irregular movements or irruptions have attracted considerable interest (Newton 1972, 2006; Bock and Lepthien 1976; Koenig and Knops 2001; Summers et al. 2024). Irruptions come with relying on seeds whose production fluctuates synchronously over large areas (Chapter 2).

Conifers other than ecotype-specific key conifers can also support crossbill populations for one or more years. For example, individuals of the Ponderosa Pine ecotype are often found in the Sierra Nevada of California during years when Lodgepole Pine in the range produces large seed crops. Unlike in the Rocky Mountains, Lodgepole Pine in the Sierra (*Pinus contorta* var. *murrayana*, a different variety from that in the Rocky Mountains) is not serotinous: its relatively soft cones open in autumn and do not hold seeds reliably for extended periods of time. I therefore doubt that there is a self-sustaining population of crossbills in the Sierra Nevada and suspect their occurrence there would cease without periodic influxes of crossbills from elsewhere. Similarly, I believe the occurrence of the Western Hemlock ecotype and other infrequent call types in eastern North America are reliant on irruptions from the west. Yet this may not always be the case: if native or exotic conifer forests in the east grow and mature, crossbills of one or more call types may establish self-sustaining populations—just as Red Crossbills have in Britain.

In contrast to North American crossbills that occur throughout the range of their key conifer and often beyond, the largest of all crossbills—the Parrot Crossbill—generally occurs in only a small fraction of the range of the Scots Pine it has specialized on (Newton 1972). Although I have spent relatively little time within the range of the Parrot Crossbill, I was fortunate to observe a pair foraging quietly on closed Scots Pine cones in September 2007. I was staying with Alan Knox in Aberdeen at the time, who had also noted the limited range of the species (Knox 1990). We met up with Mick Marquiss, a local ornithologist who led us to some locations along the River Dee where he was studying Parrot Crossbills. There we found them. This was not surprising, as Mick has spent a considerable time in the field studying crossbills, and is field-savvy. As the pair of Parrot Crossbills foraged above, we watched them use their powerful bills to separate the closed scales of the tree's cones and extract their seeds. When finished, the cones fell near our feet, littering the ground (Fig. 9.9).

Parrot Crossbills occur regularly in northern Europe east to the Ural Mountains, but rarely elsewhere within the vast range of Scots Pine, the pine with the largest natural distribution in the world (Fig. 9.10). During the last few decades, Parrot Crossbills have also established a small population in Scotland (Summers 2002),

FIGURE 9.9 A Parrot Crossbill *Loxia pytyopsittacus* foraging on a Scots Pine *Pinus sylvestris* cone (left), and two Scots Pine cones that had just been dropped by foraging Parrot Crossbills on 12 September 2007 in Ballater, Scotland (right). Their massive bills enabled them to bite seemingly effortlessly between the scales to access the seeds. Parrot Crossbill photograph taken by Philip Francis Thomsen in Asserbo, Denmark on 6 November 2021.

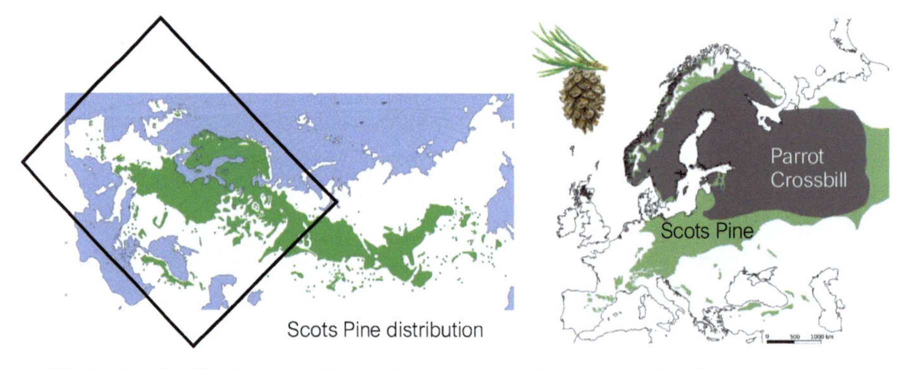

FIGURE 9.10 The distribution of Scots Pine *Pinus sylvestris* on the left, with the area outlined in a rectangle expanded on the right showing the distribution of Parrot Crossbills *Loxia pytyopsittacus* in Scotland and northern Europe. From Benkman (2024).

where the three of us watched them that September day. Scotland also harbors the Scottish Crossbill *Loxia scotia*, which has historically been thought of as a Scots Pine specialist (Nethersole-Thompson 1975). Island endemics like the Scottish Crossbill typically only occur in a small part of the range of the conifer species on which they rely, a pattern shared with Parrot Crossbills. Unlike island endemics, however, large barriers (e.g., ocean) do not isolate Parrot Crossbills from a vast majority of the range of Scots Pine.

Geographic variation in the structure of cones is a potential explanation for crossbills occurring in only a fraction of a conifer's distribution. An example is the Cassia Crossbill *Loxia sinesciuris* discussed earlier. It is restricted to 67 km^2 of the more than 200,000 km^2 that Rocky Mountain Lodgepole Pine *Pinus contorta* var. *latifolia* dominates. The Cassia Crossbill is locally adapted to foraging on the large, thick-scaled serotinous cones found within its small range. Although there is geographic variation in the structure of Scots Pine cones, I have been unable to discern any pattern in this variation coincident with the distribution of Parrot Crossbills.

Geographic variation in seed shedding phenology is another potential factor limiting the distributions of crossbills. As mentioned in Chapter 4, 'Predicting and testing specialization', the cool, wet conditions in the Pacific Northwest might account for why the western subspecies of Douglas-fir more reliably holds seeds in its cones than does the Rocky Mountain subspecies, and therefore why the Douglas-fir ecotype is more common within the range of the former subspecies than the latter. I found the idea of testing this hypothesis for Parrot Crossbills especially appealing.

An opportunity to do so was provided by Ron Summers, who had facilitated my two previous trips to Scotland. Ron had published on Scots Pine seed shedding phenology (Summers and Proctor 2005), accumulating seed fall data for a total of 16 years across four sites. Ron lives in Inverness, Scotland and has studied crossbills in Scots Pine woodlands for over 20 years, publishing numerous outstanding papers on crossbills. He and his wife, Bozena Kalejta, graciously hosted me when I was in Inverness. Ron arranged for me to stay in a rustic lodge in Abernathy Forest, the location of most of Ron's research. Located in a large 'ancient' Scots Pine woodland owned by Ron's employer (Ron is now retired), the Royal Society for the Protection of Birds, the lodge was west of where I watched Parrot Crossbills with Alan and Mick.

Seed fall data had also been published for Scots Pine in central Sweden (Hannerz et al. 2002), another region within the range of Parrot Crossbills (Fig. 9.10). What we needed, then, was seed fall data from outside the range of the Parrot Crossbill to contrast with those from the colder, more humid north. Eduardo Mezquida took on the task and led the resulting study. Conveniently located in Madrid, Eduardo could measure seed fall outside the range of Parrot Crossbills, doing so for four sites in northern Spain over three years. His analytical and intrepid field skills led to the success of the project, and to the success of field studies on crossbills and Aleppo Pine, Mountain Pine *Pinus uncinata* and Scots Pine.

Eduardo used the seed fall data to estimate when half the seeds had fallen at each site in each year, and then correlated these estimates to site- and year-specific temperature and precipitation data. He chose to use the point at which half the seeds had fallen; other thresholds did not alter his conclusions. The best predictor of this 50% seed fall benchmark was the maximum temperature in April each year. The close negative correlation between these variables is shown in Figure 9.11: Seeds fell earlier when temperatures were higher. Similar relationships occurred for March maximum temperature and precipitation: The higher the maximum temperature and the lower the precipitation in March, the earlier seeds fell. Seed fall occurred on average 1.5 to 2 months earlier in Spain than in Scotland or Sweden.

These data suggest a mechanism for the southern range limit of Parrot Crossbills in Europe. Based on our graphical model of seed profitability from

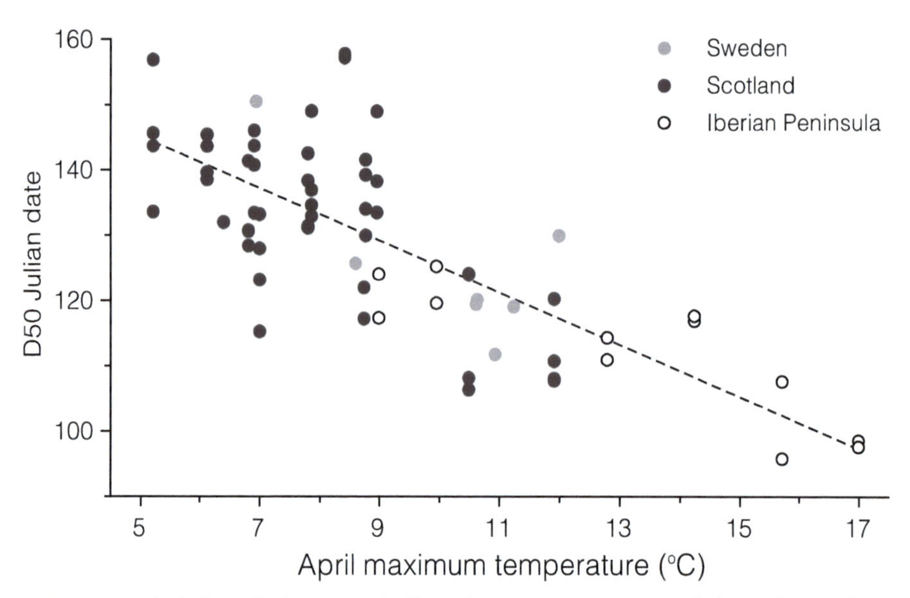

FIGURE 9.11 Relationship between April maximum temperature and date when 50% of the Scots Pine *Pinus sylvestris* seeds were released (D50) for 67 year/forest combinations in three regions. Day 120 corresponds to 30 April. Modified from Mezquida et al. (2018).

Chapter 2, 'Seed ripening and shedding' (Fig. 2.2), earlier shedding of seeds (Fig. 9.12B) results in a longer window of food scarcity, preventing crossbills from relying on Scots Pine. Farther north in Scotland and Scandinavia, seed fall is delayed, with the consequence that intake rates on Scots Pine remain sufficient into July, at which point the next seed crop becomes increasingly profitable (Fig. 9.12A).

Eduardo then used the distribution of Scots Pine combined with data on climatic variation (March and April maximum temperature, March precipitation) to predict where Parrot Crossbills are found. To determine how these variables were correlated with Parrot Crossbill occurrence, he used observations of Parrot Crossbills from breeding bird surveys during May and June. Though this is a bit later than when Parrot Crossbills usually nest, it is when Scots Pine seed would be least available (Fig. 9.12) and thus is potentially the most critical and limiting period of the year. Compellingly, maximum April temperature was a strong (and the most important) predictor of Parrot Crossbill occupancy in the model. Clearly, Scots Pine is essential, but not sufficient. Parrot Crossbills occurred almost exclusively in the portion of the pine's overall distribution where temperatures one to two months earlier in April were colder, and thus where seeds should have been held within cones longer. Areas of Scots Pine with warmer, more arid conditions are less likely to support crossbills than areas with cooler, more humid conditions, and this presumably explains why a Scots Pine specialist has such a restricted distribution.

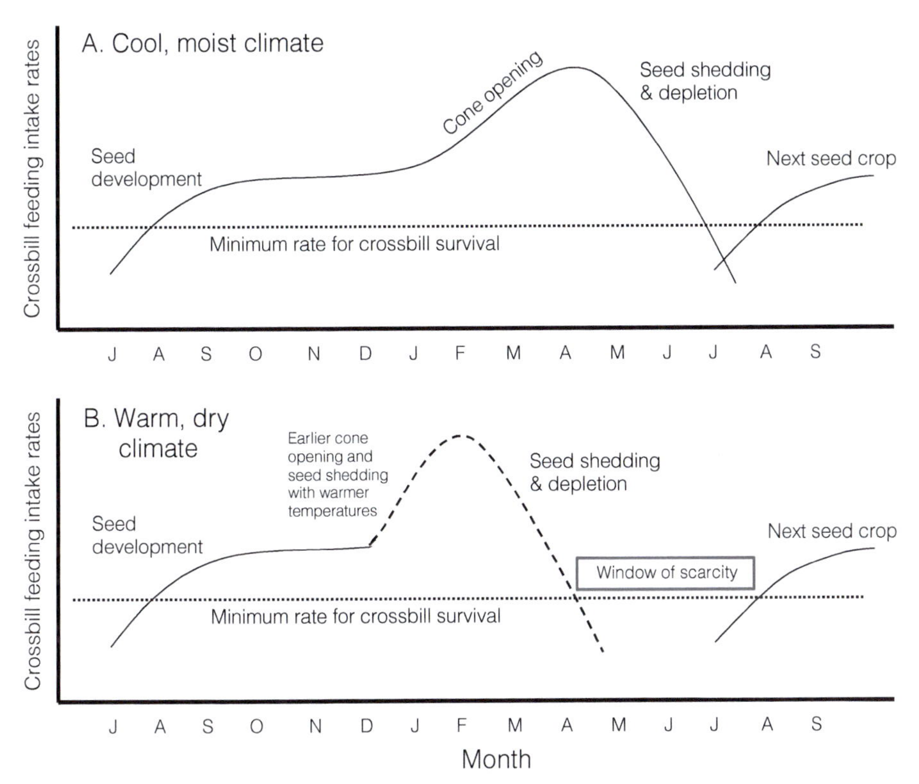

FIGURE 9.12 Seasonal patterns of seed profitability for crossbills foraging on Scots Pine *Pinus sylvestris* cones. The period of greatest food scarcity occurs between the shedding of seeds and the maturation of the next seed crop. When conditions are hotter and drier (B), seeds are shed more rapidly, and the window of food scarcity increases. The necessary seed intake rate for survival is shown as a horizontal line, although it varies seasonally with time available for foraging and daily energy demands (see for example Fig. 2.7 for seasonal variation in the necessary seed intake rates for reproduction). Modified from Mezquida et al. (2018).

More generally, these results provide strong support for the hypothesis that crossbills will only specialize on conifers that hold seeds through spring and even into summer (Chapter 4), a trait that in turn can limit their occurrence. Moreover, warm, dry conditions may often be inimical for crossbills, even reducing the suitability of mature forests. In light of this, Eduardo also modeled the distribution of Parrot Crossbills in 2080 under future climate projections. He found that the distribution of Parrot Crossbills is likely to contract, shift northward and decline overall by over half (Mezquida et al. 2018), suggesting the total forest area suitable for crossbills will drop in tandem. Because this approach does not take into consideration climate-driven disturbance events like large fires, its projections likely underestimate the true reduction in crossbill populations and habitat. This is sobering and makes my memories of watching Parrot Crossbills even more precious.

Prospects and hope

The outlook for many crossbill populations is discouraging: Mature, cone-bearing conifer forests are likely to decrease in areal extent as well as become less suitable for crossbills. But crossbills are not alone in facing daunting threats. Many species are declining, and all too many are predicted to go extinct in the next century (Ceballos et al. 2017; Urban 2024). I worry that multiple lineages of crossbills will go extinct, including one that is rather dear to my heart—the Cassia Crossbill. However, if crossbills can persist over the next century or two, and cooler climates eventually return, I am optimistic that crossbills will rebound and radiate to exploit any distinct key conifers that become available. It is imperative that we do everything we can to reverse our current climate trajectory and ensure we have cool, vast, geographically diverse conifer forests in which crossbills can evolve, coevolve and even re-evolve lost lineages.

Chapter 10

Epilogue

The reasons I was drawn to crossbills in the first place—the ease at which their foraging behavior can be observed, and the ease at which seed intake rates are quantified in the field and in aviaries—underlie much of what I've discussed in this book. The centrality of these consumer–resource interactions to crossbill ecology and behavior has allowed us to probe a diversity of questions, including those at the nexus of the fields mentioned by Bill Brown Jr. (1957) in the epigraph at the beginning of Chapter 1. Furthermore, population declines from climate change are generally mediated through consumer–resource or trophic interactions (Cahill et al. 2013; Ockendon et al. 2014; Pearce-Higgins and Green 2014; Renner et al. 2024). Thus, we've been well positioned for understanding some recent causes of crossbill declines.

There is more to learn, and I've mentioned some topics along the way that I think would be worth addressing. I am sure readers of this book will think of other questions, and I hope that those with the inclination and wherewithal address them. I suspect that with advances in technology, questions that I had either considered unanswerable with crossbills or hadn't considered at all will become tractable. Questions that can be addressed using genomic data are perhaps the lowest-hanging fruit to pick. For example, a fully resolved phylogeny of all crossbill taxa that provides estimates of the timing of lineage divergences, their population sizes and the extent of gene flow in the past would greatly enrich our understanding of crossbill diversification. Several of my collaborators are pursuing such studies, and I'm eagerly anticipating their results. To conclude this book, I would like to highlight several outstanding questions related to crossbill biology.

What delayed the evolution of crossbills?

Though crossbills appear to have only evolved in the last 1.4 million years (Chapter 1, 'Why crossed bills?'), cone-bearing conifers have been around for tens of millions of years, with pines diverging from spruce over 100 million years ago (Willyard et al. 2007). Even most extant species of pine have been present for 3 to 5 million years (Leslie et al. 2012; Jin et al. 2021; Zhao et al. 2024), long before crossbills came onto the scene. Moreover, cone-bearing conifers undoubtedly have been subject to predispersal seed predation by birds and mammals for tens of millions of years (Leslie 2011). This raises a lingering question: If, as I argued in Chapter 1, the evolution of the bill crossing did not require extraordinary circumstances, why didn't crossbills appear earlier?

The one explanation I can offer is that the evolution of crossbills is tied to the period of the greatest cooling and coldest temperatures of the last 65 million years, which commenced 1.5 to 0.9 million years ago (Zachos et al. 2001; Clark et al. 2024). As discussed previously, colder temperatures cause cones to remain closed, thus retaining seeds for longer. Before the onset of exceptionally cold temperatures during the last 1.5 million years, cones may have regularly shed most if not all their seeds before spring. This is especially true for conifers with thin-scaled cones, which were likely the most important to an ancestral crossbill that presumably had a redpoll-like bill morphology and body size. As the climate cooled, seeds in thin-scaled cones undoubtedly became more regularly available in late winter and spring. Reliable cone-seed at this time of the year—when seeds are generally most scarce and limiting for cardueline finches (Newton 1967, 1972)—may have been the critical resource enabling a relatively small ancestral crossbill with slightly crossed mandibles to thrive. Once a population of such birds became established, the evolution of a greater mandible crossing and associated adaptations furthering the exploitation of cone-seed would seem almost inevitable (Chapter 1, 'Why crossed bills?').

While the relatively cold temperatures of the last 1.5 million years may have been what enabled an ancestral crossbill to specialize on conifers, greater seasonality, especially at higher latitudes (Zachos et al. 2001) where crossbills presumably evolved, may have also been important. Increasing seasonality leads to greater resource partitioning in birds, likely because of increasing resource competition (Porter et al. 2022). Indeed, such a mechanism could help explain why bill shape diverges more rapidly at higher latitudes (Freeman et al. 2022). Perhaps what favored the evolution of crossbills was the combination 1.5 million years ago of increasing seasonality that increased food competition among forest passerines—selecting for reliance on alternative resources—and colder temperatures causing cone-seed to become a reliable alternative resource during the leanest times of the year.

The unfortunate implication of this hypothesis, as discussed in Chapter 9, is that ongoing and accelerating climate change will cause populations of crossbills to decline and possibly go extinct. The adverse effects of warmer temperatures on the availability of cone-seed are further supported by demographic reconstructions of the effective population sizes (N_e) of White-winged Crossbills *Loxia l. leucoptera* and Red Crossbills *L. curvirostra* from North America using whole-genomes (Germain et al. 2023). The population sizes of both white-wings and reds decreased when temperatures increased sharply by 8°C between approximately 147,000 and 123,000 years ago. Conversely, their population sizes increased when average global surface temperature subsequently declined by 8°C between approximately

122,000 to 65,000 years ago, consistent with the importance of colder temperatures in increasing the reliability of cone-seed to crossbills.

What about crossbill diversity elsewhere?

An even greater diversity within Red Crossbills—both in range of bill and body sizes and number of subspecies—has been described in the Palearctic than in North America (Griscom 1937; Newton 1972). Most of the described subspecies are geographically isolated on the periphery of the large range of the species (Newton 1972). Some of these peripheral subspecies are clearly associated with a single species of conifer (e.g., both *L. c. poliogyna* in northwest Africa and *L. c. balearica* on Mallorca rely on Aleppo Pine *Pinus halepensis*; *L. c. meridionalis* in Vietnam and *L. c. luzoniensis* in the Philippines rely on *P. kesiya*), and for some there are plausible conifers on which they might specialize (e.g., *L. c. himalayensis* and *L. c. bangsi* in the Himalayan region [Edelaar 2008]). In mainland Europe, however, the situation is less clear. Here, one subspecies (the Common Crossbill *L. c. curvirostra*) is currently recognized, which apparently comprises more than 15 call types (Robb 2000; Martin et al. 2019). Although there is greater morphological variation within the Common Crossbill than is found within other crossbill subspecies (Edelaar and Terpstra 2004), it is doubtful that each call type represents a different specialized ecotype, for the simple reason that there are not that many suitable conifer species in Europe (Martin et al. 2020). Indeed, some call types differ little in bill size (e.g., Summers 2020), and individuals of some populations recorded in western Europe likely reside mostly elsewhere (e.g., Newton 2006; Marquiss et al. 2012). Moreover, work in Europe indicates that call types might evolve quickly and go extinct nearly as fast (Martin et al. 2024), possibly as a result of rapid cultural evolution without divergent ecological adaptation. Regardless, making sense of this variation within the Common Crossbill will require geographically extensive surveys of vocalizations (e.g., Martin et al. 2019) combined with morphological measurements of the same individuals (e.g., Groth 1993a), along with focused studies on behavior, ecology and evolution like those I have discussed in this book.

In Spain, where more comprehensive studies have been undertaken, different populations of crossbills are characterized by distinct contact calls, bill sizes and shapes. Each is associated with (and appears adapted to) different species of pine (Edelaar et al. 2012; Alonso et al. 2020). Extensive studies in Scotland have revealed much about the behavior and ecology of three species of crossbills—Common, Scottish *Loxia scotia* and Parrot *L. pytyopsittacus*—including their seasonal and differential use of conifer species (Marquiss and Rae 1994, 2002; Summers et al. 2002, 2010). Especially noteworthy is the extensive reliance on non-native conifers by Common and Scottish Crossbills (Marquiss and Rae 1994; Summers et al. 2002; Summers 2018). Moreover, the stand structure of the sole native cone-bearing conifer in Scotland, Scots Pine *Pinus sylvestris*, has apparently shifted to younger, more closed forests, which have larger cones (see Summers and Proctor 1999).

This may have favored the establishment and increase of the larger-billed Parrot Crossbill over the endemic Scottish Crossbill during the past few decades. The work also hints at the possibility of novel forms of selection, and foregrounds the challenge of understanding the causes of crossbill diversification in highly altered environments (Edelaar et al. 2012), and in areas where they may be simultaneously adapting to more than one conifer species. Indeed, the relative simplicity of studying crossbills in areas dominated by vast tracts of relatively undisturbed native conifers (like eastern Canada and western North America), combined with aviary experiments to test specific hypotheses, has been critical for understanding North American crossbill ecology and diversity.

A striking contrast worthy of further study

Although variation in the extent of coevolution is responsible for slight to substantial differences among crossbill populations (Figs 5.1 and 5.2), most of their morphological variation is attributable to adaptation to alternative conifers (Chapter 4). This is well illustrated by the massive size difference between the largest and smallest crossbills, the Parrot and a Himalayan subspecies of Red Crossbill (*L. c. himalayensis*), respectively (Fig. 10.1A). These taxa rely on structurally different cones from different genera of conifers. Yet it was the nearly as extreme differences between the two subspecies of Red Crossbill that rely on the same species of pine, *Pinus kesiya* (Fig. 10.1), which caused me to wonder whether coevolution could drive ecotypic divergence without this obvious source of selection.

The larger of these two subspecies, *L. c. meridionalis*, occurs in a small, isolated area of *P. kesiya* on the Langbian Plateau in southern Vietnam (Fig. 10.1B; Griscom 1937; Clouet and Goar 1999). This Vietnam Crossbill (Fig. 10.1A) is the second-largest crossbill (12.37 mm mean bill depth [mean of two sexes] versus 13.66 mm for the Parrot Crossbill). The smaller of the *P. kesiya* specialists, the Philippine Crossbill *L. c. luzoniensis* (Fig. 10.1A), is confined to *P. kesiya* forests in the mountains of northwestern Luzon, Philippines (Fig. 10.1B; Griscom 1937; Clouet and Goar 2001), and is among the smallest of crossbills (8.47 mm mean bill depth versus 7.74 mm for *L. c. himalayensis*). The difference in mean bill depth between these two *P. kesiya* specialists (3.90 mm) represents two-thirds of the total range (5.92 mm) found between all crossbill taxa (Fig. 10.1A), and exceeds the total range (2.80 mm) found between all New World Red Crossbills. This difference is seven and a half times greater than the mean bill depth difference between the Cassia Crossbill *L. sinesciuris* and the other Lodgepole Pine *P. contorta* specialist (Type 5; 0.52 mm).

Why is there such a great difference between the two *P. kesiya* specialists? One possibility is variation in the degree of coevolution between crossbills and pines. *Pinus kesiya* has likely occurred in southern Vietnam (on the Langbian Plateau) for hundreds of thousands of years and apparently spread, also perhaps

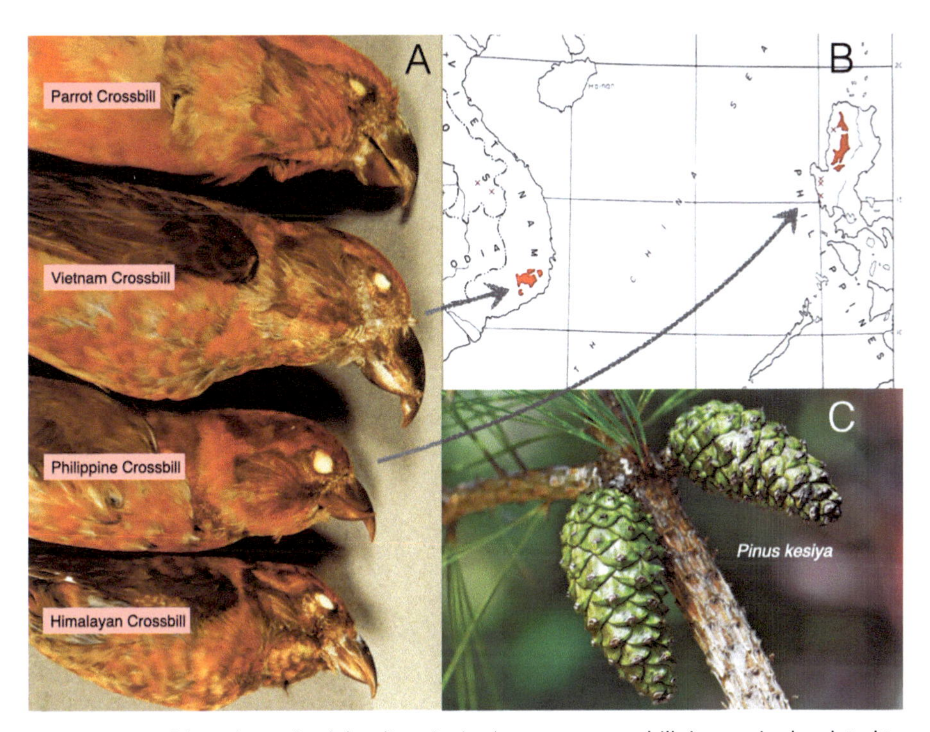

FIGURE 10.1 Although much of the diversity in size among crossbills is seemingly related to the diversity of cone structures of the different conifer species on which crossbills specialize, the potential for coevolution alone to contribute to crossbill diversity is hinted at by two crossbill subspecies that rely almost exclusively on seeds in the cones of the same species of pine, *Pinus kesiya* (C). The Vietnamese Crossbill *Loxia c. meridionalis* (A) endemic to the Langbian Plateau in Vietnam (B) is the second-largest crossbill, larger than all but the Parrot Crossbill *L. pytyopsittacus*. The Philippine Crossbill *L. c. luzoniensis* (A), meanwhile, is endemic to the Central Cordillera on the island of Luzon in the Philippines (B) and is among the smallest of crossbills. The range of their average bill depths represents two-thirds of the total range among all crossbills (A). A: specimens photographed at the National Museum of Natural History, Washington, DC; B: range map of *P. kesiya* from Critchfield and Little (1966); C: cones photographed near Baguio City, Philippines on 6 December 2010.

hundreds of thousands of years ago, to the Philippines during a glacial period in the Pleistocene via the Sunda Shelf, Borneo and Palawan (Mirov 1967; Gao et al. 2024). Populations of *P. kesiya*, however, have undoubtedly been consistently larger in Vietnam than in the Philippines (Turnbull et al. 1980; Werner 1997; Penny 2001; Gao et al. 2024), where *P. kesiya* was sparse and became common only during the last 1,000 years (Stevenson et al. 2010). We thus might expect that a coevolutionary arms race between crossbills and pine has escalated to a much greater extent in Vietnam than in the Philippines, potentially explaining the evolution of large bills in Vietnam but not the Philippines.

Though the massive bill of the Vietnam Crossbill is consistent with greater coevolutionary escalation, differences in cone traits such as scale thickness between the two regions would provide stronger evidence for this hypothesis. Michel Clouet, a French ornithologist who piqued my interest in Mountain Pine *Pinus uncinata* in the Pyrenees during a visit to my home in the 1990s, measured cones in both areas and found a trend for larger cones and scales in Vietnam than in the Philippines (Clouet 2003). However, as relatively few cones were measured from an unknown number of trees, I felt a need for larger samples and measures of additional cone traits. Seeing both crossbills and pines firsthand would also be valuable. Thus, Tom Parchman and I traveled to Vietnam and the Philippines in December 2010. Our goal was to measure selection exerted by crossbills and test whether there were consistent differences reflecting greater crossbill–pine coevolution in Vietnam than in the Philippines. Critical for our work in Vietnam was Dr. Luu Hong Truong from the Institute of Tropical Biology at the Vietnam Academy of Science and Technology, who arranged all permits and local travel in addition to helping us collect cones.

In Vietnam, we observed crossbills foraging on open and partly open cones, and found shredded cones on multiple trees, which is evidence of closed-cone feeding. We did *not* find any evidence of cones being shredded in the Philippines, which would have allowed us to assess seed predation by crossbills. Our observations were consistent with those of Clouet (2003) who concluded that small-billed Philippine Crossbills were limited to feeding on seeds between open scales.

In total, we were able to gather two to three cones from 20 or more trees at three locations in each country. The cones were not noticeably different between Vietnam and the Philippines, and I was surprised that we did not detect any difference in the thickness of the cone scales. In addition, a tree we sampled in Vietnam that crossbills clearly favored did not have relatively thin scales. Cone mass and length didn't seem to be important, either. Instead, cones in Vietnam were characterized by features that Chris Smith (1970) argued were defenses directed at tree squirrels—a low ratio of seed mass to cone mass, and higher cone density (or harder cone scales). Moreover, observations of foraging crossbills in Vietnam indicate that these traits challenge and deter crossbills: the same intensively harvested tree had a relatively low ratio of cone mass to seed mass and relatively less dense or softer scales.

Conceivably, selection by tree squirrels in Vietnam (*Callosciurus* sp. and *Dremomys* sp., especially *C. erythraeus*; Koyabu et al. 2009) could have led to enhanced squirrel defenses that, as a byproduct, favored the evolution of the large bill of the Vietnam Crossbill (much as we have argued for Aleppo Pine and Red Crossbills on the Iberian Peninsula; Mezquida and Benkman 2005). The small bill of the Philippine Crossbill could reflect either cones with fewer defenses (tree squirrels are absent from the Philippines) or a crossbill taxon that has not yet evolved the ability to access seeds in closed cones, perhaps due to the limited time crossbills have occurred in the archipelago. This is all rather speculative—perhaps too speculative. However, I find it a fascinating system and mention it here in the hope that someone studies it further. Our findings in Vietnam and the Philippines also make me wish I had quantified scale or cone density (mass/volume) in other studies, as it is a trait that likely influences the cone choices of seed predators (Smith 1970; Greene et al. 2024).

What can crossbills tell us about other birds and ecology, and evolution more generally?

I have limited my discussion of the general implications of our work on crossbills, for two reasons. One is that it makes for a more cohesive story about crossbills themselves. Another is that those readers most interested in big-picture takeaways are likely aware that they can be found in the Introduction and Discussion sections of the scientific articles I have cited throughout this book. Nonetheless, one might question whether our findings can be generalized to other bird species, given the degree to which crossbills are specialized on cone-seed, and that they often rely on the seeds of only one to several species of conifers. Specialization on a single food species certainly helps account for why the outcome of the coevolutionary interactions between Cassia Crossbills and Lodgepole Pine is so predictable (Chapter 5). In contrast, if interactions were among a larger set of species—as is common for many other taxa—the signatures of coevolution would likely be harder to detect, if they were even present. Regardless, I believe many of the insights we have gained are not limited to crossbills and conifers (Thompson 1994, 2005, 2013).

Although relatively few bird species are as specialized as crossbills, diets dominated by one or several food species occur among a diversity of birds including several species of hummingbird (Temeles and Kress 2003; Boehm et al. 2022), both species of sage-grouse *Centrocercus* spp. (Schroeder et al. 2020; Young et al. 2020) and several species of southern Australian black-cockatoos *Calyptorhynchus* spp. (Joseph 1986; Pepper et al. 2000), to name a few. The black-cockatoos are especially intriguing because Australia is analogous to the South Hills in lacking tree squirrels and containing birds as the most important tree-seed predators (Low 2014). And serotiny is common in Australia, more common than anywhere else in the world (Lamont et al. 1991). But instead of crossbills, Australia has an abundance and great diversity of parrots, and some of that diversity owes its existence to seeds in hard, woody serotinous fruits (Low 2014). The similarities of Australia's parrots to crossbills were noted to me by the renowned tropical ecologist John Terborgh (pers. comm. November 2023). John wrote, 'Australia's parrots are crossbills writ large with a mind-numbingly diverse flora ... to exploit. Countless times, watching Australian parrots has reminded me of your work. In a second incarnation I think you would have a lovely time studying the much more complex interactions between parrots and trees in Australia.' A lovely thought indeed!

More generally, when diets are quantified across a diversity of species, one or two food species comprise over half of the diet of many populations of vertebrates (Hutchinson et al. 2022). Thus, although crossbills are near the extreme end of diet specialization among vertebrate species, a disproportionate reliance on one or a few prey species is likely more common than generally recognized. If we are to include the diets of insects—especially the vast number of phytophagous (plant-eating) insect species—extreme diet specialization (e.g., consuming only one or a few closely related species) is often the norm (Forister et al. 2015). Consequently, the interactions between crossbills and conifers are likely more broadly representative of consumer–resource interactions in nature than those found in most bird species.

To demonstrate the broader relevance of our work, I will highlight some general implications of our study on variation in reproductive isolation and its causes, first discussed in Chapter 8. You will recall that Cody Porter demonstrated that different components of reproductive isolation increased between two ecotypes as their differences in intake rates increased on shared conifers (Figs 8.3 and 8.4), providing a compelling example of ecological speciation (Schluter 2009; Nosil 2012). Although there are numerous other examples of this process across the tree of life, few have used quantitative measures of divergent selection, and most are devoid of much ecological context (Nosil 2012). Indeed, it should be clear by now that being able to quantify consumer–resource relationships is often the main (and in some cases the only!) advantage of studying crossbills. Yet the point I want to emphasize is that by comparing measures of reproductive isolation across a range of foraging tradeoffs (Fig. 8.3), Cody could show that reproductive isolation is very strong to complete when foraging tradeoffs are strong, but weak when tradeoffs are weak and food is plentiful.

We believe a broader takeaway from this result is this: because the diets of most breeding birds converge on plentiful food like caterpillars for which foraging tradeoffs are most likely weak (Porter et al. 2022), birds are unlikely to speciate in sympatry (Porter and Benkman 2022). (My favorite example of trophically dissimilar birds converging on a common resource when nesting are the massive-billed Evening Grosbeak *Hesperiphona vespertina* and the much smaller and slender-billed Bay-breasted Warbler *Setophaga castanea*, both of which feed on Spruce Budworms *Choristoneura fumiferana* when nesting in the North American boreal forest.) Thus, as a general rule, birds may need to evolve in allopatry for speciation to occur. Of course, this is hardly a novel insight (Mayr 1963; Price 2008). However, I think the value of Cody's study is that it provides evidence of a general mechanism underlying a well-documented pattern.

Finally, I am excited by the potential of fecal DNA metabarcoding to improve our understanding of the diets and ecology of hard-to-study species. One recent example of this approach involves Zebras *Equus quagga*, Wildebeests *Connochaetes taurinus* and Thomson's Gazelles *Eudorcas thomsonii*, where previously unknown dietary preferences helped unravel the much-debated causes of sequential migration by these species across the Serengeti (Anderson et al. 2024). Indeed, when describing the insights gained from using various methods including fecal DNA metabarcoding, Robert Pringle (2023) wrote: 'The seemingly messy details are often not so messy when the right tools are brought to the job.' I feel very fortunate that I began studying crossbills when the 'right tools' were basic and available: a telescope, a stopwatch, calipers and a scale. Given that virtually all my insights on crossbills ultimately came from measures of intake rates, I'm reminded of the quote by G. Evelyn Hutchinson that I included in Chapter 2, 'Seed ripening and shedding'. Hutchinson (1959: 147) wrote:

> In any study of evolutionary ecology, food relations appear as one of the most important aspects of the system of animate nature. There is quite obviously much more to living communities than the raw dictum 'eat or be eaten,' but in order to understand the higher intricacies of any ecological system, it is most easy to start from this crudely simple point of view.

We can learn a lot without knowing much about food relations, but behavioral, ecological and evolutionary studies are often richer and more persuasive when food relations are more than an afterthought or black box. If you are still not convinced, let me conclude by mentioning two of arguably the most influential studies in evolutionary biology since I started studying crossbills. The first is the 40-year study on Darwin's finches led by Peter and Rosemary Grant; the second is the 30-plus-year study of stickleback led by Dolph Schluter. Justifiably famous, both studies were built upon detailed observational and experimental studies, respectively, on 'food relations'. Without that foundation, evolutionary biology would be much poorer.

Glossary

Adaptation: The process by which a population becomes better suited to its environment by natural selection or behavior such as performance-based habitat choice.

Allometric relationship: A quantitative relationship between a feature of an organism (e.g., bill size or daily energy requirements) and its body size.

Allopatric: Refers to populations or species occupying separate geographic areas.

Antisymmetry: An asymmetric trait (e.g., mandible crossing direction, handedness) that occurs in equal proportions (1:1) in a population.

Assortative mating: Mating is nonrandom based on some trait such as size, color or vocalizations. Individuals mating with others sharing their traits (e.g., large-billed birds mating with other large-billed birds) is referred to as positive assortative mating. Because negative assortative mating (e.g., birds preferring to pair with individuals with dissimilar contact calls over those of their own call type) is not relevant to the examples in this book, I simply refer to assortative mating assuming positive assortative mating.

Behavioral isolation: A premating reproductive isolating barrier that arises because of behavior that inhibits individuals from courting or copulating.

Call type: Populations of Red Crossbills that are characterized by distinctive contact calls (and other vocalizations) and whose morphological variation is comparable to that found in many species of songbirds.

Call switching: When individuals modify their contact calls such that they conform to those of a different call type. This has not been detected in captive crossbills, but it has been detected albeit rarely in a banded population of Cassia Crossbills and in one individual of the Ponderosa Pine ecotype.

Canopy seed bank: Seeds that accumulate in the canopy of woody plant species such as in Rocky Mountain lodgepole pine.

Carrying capacity: The number of individuals in a population that can be supported by resources in the habitat.

Character displacement: Adaptive evolution (generally divergence) of characters (e.g., bill and call structure) as the result of negative interactions between species.

Coevolution: Reciprocal natural selection between two entities—usually two species—that causes reciprocal evolution in the two entities.

Coevolutionary arms race: The ongoing escalation of defense and counter defense as the result of coevolution between species.

Coevolutionary diversification: Coevolution that causes lineages to diversify into multiple taxa or species.

Condition-dependent mate choice: Mate choice that is dependent on the condition of perspective mates.

Contact call: A brief, commonly used vocalization that coordinates activities in crossbills (e.g., attract flockmates, coordinate flock movements). Also referred to as a flight call, however such calls are often given when crossbills are perched and not necessarily associated with flight.

Cultural evolution: Intergenerational change in socially transmitted behavior.

Directional asymmetry: An asymmetry that is unvarying in individuals in a population (e.g., all Wrybills have bills that curve to the right).

Ecological speciation: When reproductive isolation arises between populations because of their adaptation to alternative environments. In the case of crossbills, reproductive isolation arises between populations because of their adaptation to alternative species or populations of conifers.

Ecotype: A genetically controlled phenotype of a species associated with particular resources or environmental conditions. I refer to those call types of Red Crossbills for which there is evidence of adaptation to particular conifers as ecotypes.

Exploitative competition: Competition between individuals that arises from their depletion of shared resources.

Extrapair paternity: Paternity that arises from copulations outside the social pair.

Fitness surface: A plot of how the expected fitness of an individual varies in relation to changes in phenotype (observable traits such as bill size, husking groove width).

Forest regeneration: Recovery of forests following disturbance such as high-intensity, stand-replacing fire.

Geographic mosaic of coevolution: The theory developed to understand how and why interspecific interactions (e.g., predator-prey, mutualisms) and resulting coevolution vary in space and among populations. Of particular interest are examples where the strength of reciprocal interactions and the outcome vary among populations (e.g., some populations experience strong reciprocal selection and coevolve, whereas others don't), and where the outcomes are altered by migration and gene flow or other factors.

Habitat isolation: A premating reproductive isolating barrier that arises when individuals of different groups occupy different habitats preventing their interbreeding.

Heritability: The extent to which variation in a trait in a population is the result of genetic factors.

Host switch: Generally refers to situations where a parasite evolves to utilize a new host species. Ecotypes of Red Crossbills are much like host races of phytophagous insects that specialize on different species of plants (or feather lice that specialize on different species of birds). And like host races of insects, the process by which populations of crossbills adapt to an alternative plant species can be referred to as a host switch.

Immaturity risk: When regeneration is unlikely because disturbances (i.e., stand-replacing fire) occur a plant population reaches reproductive maturity or before a canopy seed bank develops in a serotinous population like Rocky Mountain Lodgepole Pine.

Immigrant inviability: A form of premating reproductive isolation where immigrants are unable to survive or have relatively low survival compared to residents (i.e., those locally adapted), with the result that immigrants are prevented from interbreeding with residents. This is a particularly important reproductive isolating barrier for species that have limited or no capacity to differentially settle among habitats.

Liem's Paradox: When a species has a morphology, behavior and physiology that indicates specialization for feeding on a particular prey type, yet the species sometimes or even often feeds generally.

Local adaptation: Individuals have higher fitness (greater survival and/or reproduction) in their 'home' environment than elsewhere. For example, the Ponderosa Pine ecotype of Red Crossbill has higher feeding rates and a greater likelihood of breeding in Rocky Mountain Ponderosa Pine forests than in nearby Rocky Mountain Lodgepole Pine forests, and vice versa for the Lodgepole Pine ecotype.

Mandible crossing direction: The side to which the lower mandible twists and crosses the upper mandible.

Marginal value theorem: A model used for determining the optimal time to spend in a patch (e.g., a food patch such as a tree or conifer cone).

Masting: Occurs when a plant population produces large annual seed crops interspersed between years when relatively few or no seeds are produced.

Matching habitat choice: see Performance-based habitat choice

Natural selection: Differential survival and/or reproduction of individuals in a population based on differences in one or more traits.

Negative frequency-dependent selection: A mode of natural selection in which the fitness of a particular type (a genotype such as a particular gene combination or phenotype such as crossbills whose lower mandibles cross to the right) decreases with increases in its frequency in the population. Sex ratio is a good example, where the expected reproductive success of for example a male decreases as the frequency of males increases in the population.

Nomadism: Individuals move from place to place periodically or seasonally but need not and perhaps rarely return to prior places.

Palate, horny: The lower or inner surface of a bird's upper mandible.

Performance differential: The standardized difference between the mean feeding intake rates of two taxa (e.g., ecotypes) of crossbills. It is a quotient that provides a measure of proportionately how much less efficient one taxa or ecotype is relative to the other.

Performance-based habitat choice: Individuals preferentially occupy habitats where their performance (e.g., intake rate, survival, reproduction) is greatest.

Phenology: The study of the timing or seasonality of events.

Phenotype: Includes all observable traits (e.g., morphological, physiological, behavioral) of an organism.

Postzygotic reproductive isolation: Reproductive isolation that occurs following fertilization; includes hybrid inviability and sterility, and poor performance by hybrids in the environment.

Profitability: The value of a food type (e.g., calories) per unit foraging time.

Reduced immigrant fecundity: A form of premating reproductive isolation where immigrants are less able to reproduce compared to residents (i.e., those locally adapted), with the result that immigrants are prevented from interbreeding with residents.

Reproductive isolation: Reduction of gene flow between populations; see various Reproductive isolating barriers and Postzygotic reproductive isolation.

Reproductive isolating barrier: see Behavioral reproductive isolation; Habitat isolation; Immigrant inviability; Reduced immigrant fecundity

Serotiny: Retention of seeds in woody fruits for a year or more. The cones of many Rocky Mountain Lodgepole Pine trees remain closed for years and even decades. Cones open and shed seeds after being heated by fire, very hot days or extended weathering. Serotiny is considered an adaptation to high-intensity, stand-replacing fires that are likely to occur within the expected lifetime of the woody plant.

Sibling species: Species that are indistinguishable morphologically, but differ in other ways (e.g., behavior, genetics) that indicate they are reproductively isolated.

Sister species: Pairs of species that are each other's closest relatives.

Speciation: The evolution of reproductive isolation within a species producing two or more species.

Subspecies: A geographically distinct population or group of populations that is geographically isolated from other subspecies. Of critical importance is exclusive breeding ranges; nonbreeding ranges often overlap in migratory species.

Sympatric: Refers to populations or species occupying the same geographic area.

Taxa: Taxonomic units to which related groups of individuals or species are assigned. Examples of different taxa include the subspecies White-winged Crossbill, the species Cassia Crossbill, the subfamily Carduelinae and the family Fringillidae. Taxon is the singular of taxa.

Tomium: The sharp or cutting edge of the bill.

Tradeoff: When a change in a trait or set of traits improves performance in one task while decreasing performance in another task. For example, an increase in bill size might provide greater access to seeds in large cones but decrease their efficiency at utilizing seeds in small cones. Tradeoffs also occur because of limited resources/time, and allocating resources/time to one tissue or activity (e.g., reproductive organs or provisioning offspring) reduces the availability of resources to other tissues or activities (e.g., individual maintenance).

References

Abaimov, A.P., Lesinski, J.A., Martinsson, O. and Milyutin, L. (1998) *Variability and Ecology of Siberian Larch Species*, Vol. 43. Department of Silviculture, Swedish University of Agricultural Sciences.

Abatzoglou, J.T. and Williams, A.P. (2016) Impact of anthropogenic climate change on wildfire across western US forests. *Proceedings of the National Academy of Sciences USA* 113: 11770–5. https://doi.org/10.1073/pnas.1607171113

Agne, M.C., Fontaine, J.B., Enright, N.J. and Harvey, B.J. (2022) Fire interval and post-fire climate effects on serotinous forest resilience. *Fire Ecology* 18: 22. https://doi.org/10.1186/s42408-022-00145-4

Agrawal, A. (2017) *Monarchs and Milkweed: A Migrating Butterfly, a Poisonous Plant, and their Remarkable Story of Coevolution*. Princeton, NJ: Princeton University Press. https://doi.org/10.1515/9781400884766

Aigner, P.A. (2001) Optimality modeling and fitness trade-offs: when should plants become pollinator specialists? *Oikos* 95: 177–84. https://doi.org/10.1034/j.1600-0706.2001.950121.x

Alexander, M.E. and Cruz, M.G. (2012) Modelling the effects of surface and crown fire behaviour on serotinous cone opening in Jack Pine and Lodgepole Pine forests. *International Journal of Wildland Fire* 21: 709–21. https://doi.org/10.1071/WF11153

Alizadeh, M.R., Abatzoglou, J.T., Adamowski, J., Rad, A.M., AghaKouchak, A., Pausata, F.S.R. and Sadegh, M. (2023) Elevation-dependent intensification of fire danger in the western United States. *Nature Communications* 14: 1773. https://doi.org/10.1038/s41467-023-37311-4

Allen, C.D., Macalady, A.K., Chenchouni, H., Bachelet, D., McDowell, N., Vennetier, M., Kitzberger, I., Rigling, A., Breshears, D.D., Hogg, E.H.T., Gonzalez, P., Fensham, R., Zhang, Z., Castro, J., Demidova, N., Lim, J.-H., Allard, G., Running, S.W., Semerci, A. and Cobb, N. (2010) A global overview of drought and heat-induced tree mortality reveals emerging climate change risks for forests. *Forest Ecology and Management* 259: 660–84. https://doi.org/10.1016/j.foreco.2009.09.001

Alonso, D., Fernández-Eslava, B., Edelaar, P. and Arizaga, J. (2020) Morphological divergence among Spanish Common Crossbill populations and adaptations to different pine species. *Ibis* 162: 1279–91. https://doi.org/10.1111/ibi.12835

Althoff, D.M., Segraves, K.A. and Johnson, M.T.J. (2014) Testing for coevolutionary diversification: linking pattern with process. *Trends in Ecology and Evolution* 29: 82–9. https://doi.org/10.1016/j.tree.2013.11.003

Anderson, T.M., Hepler, S.A., Holdo, R.M., Donaldson, J.E., Erhardt, R.J., Hopcraft, J.G.C., Hutchinson, M.C., Huebner, S.E., Morrison, T.A., Muday, J., Munuo, I.N., Palmer, M.S., Pansu, J., Pringle, R.M., Sketch, R. and Packer, C. (2024) Interplay of competition and facilitation in grazing succession by migrant Serengeti herbivores. *Science* 383: 782–8. https://doi.org/10.1126/science.adg0744

Andrus, R.A., Chai, R.K., Harvey, B.J., Rodman, K.C. and Veblen, T.T. (2021) Increasing rates of subalpine tree mortality linked to warmer and drier summers. *Journal of Ecology* 109: 2203–18. https://doi.org/10.1111/1365-2745.13634

Armsworth, P.R. and Roughgarden, J.E. (2008) The structure of clines with fitness-dependent dispersal. *American Naturalist* 172: 648–57. https://doi.org/10.1086/591685

Audley, J.P., Fettig, C.J., Munson, A.S., Runyon, J.B., Mortenson, L.A., Steed, B.E., Gibson, K.E., Jørgensen, C.L., McKelvey, S.R., McMillin, J.D. and Negrón, J.F. (2020) Impacts of Mountain Pine Beetle outbreaks on Lodgepole Pine forests in the Intermountain West, U.S., 2004–2019. *Forest Ecology and Management* 475: 118403. https://doi.org/10.1016/j.foreco.2020.118403

Bai, H., Sun, Y., Liu, N., Xue, F., Li, Y., Xu, S., Ye, J., Zhang, L., Chen, Y. and Chen, J. (2018) Single SNP- and pathway-based genome-wide association studies for beak deformity in chickens using high-density 600k SNP arrays. *BMC Genomics* 19: 501. https://doi.org/10.1186/s12864-018-4882-8

Bailey, A.M., Niedrach, R.J. and Bailey, A.L. (1953) *The Red Crossbills of Colorado*. Museum Pictorial Number 9. Denver: Denver Museum of Natural History.

Baker, W.L. (2006) Fire and restoration of sagebrush ecosystems. *Wildlife Society Bulletin* 34: 177–85. https://doi.org/10.2193/0091-7648(2006)34[177:FAROSE]2.0.CO;2

Baltzer, J.L., Day, N.J., Walker, X.J., Greene, D., Mack, M.C., Alexander, H.D. and Johnstone, J.F. (2021) Increasing fire and the decline of fire-adapted Black Spruce in the boreal forest. *Proceedings of the National Academy of Science USA* 118: e2024872118. https://doi.org/10.1073/pnas.2024872118

Barbero, R., Abatzoglou, J.T., Larkin, N.K., Kolden, C.A. and Stocks, B. (2015) Climate change presents increased potential for very large fires in the contiguous United States. *International Journal of Wildland Fire* 24: 892–9. https://doi.org/10.1071/WF15083

Barnosky, C.W. (1989) Postglacial vegetation and climate in the northwestern Great Plains of Montana. Quaternary Research 31: 57–73. https://doi.org/10.1016/0033-5894(89)90085-9

Behl, N.J. and Benkman, C.W. (2018) Habitat associations and abundance of a range-restricted specialist, the Cassia Crossbill (*Loxia sinesciuris*). *Condor: Ornithological Applications* 120: 666–79. https://doi.org/10.1650/CONDOR-17-257.1

Bell, D.M., Bradford, J.B. and Lauenroth, W.K. (2014) Early indicators of change: divergent climate envelopes between tree life stages imply range shifts in the western United States. *Global Ecology and Biogeography* 23: 168–80. https://doi.org/10.1111/geb.12109

Bello, C., Galetti, M., Pizo, M.A., Magnago, L.F.S., Rocha, M.F., Lima, R.A.F., Peres, C.A., Ovaskainen, O. and Jordano, P. (2015) Defaunation affects carbon storage in tropical forests. *Science Advances* 1: e1501105. https://doi.org/10.1126/sciadv.1501105

Benkman, C.W. (1987a) Food profitability and the foraging ecology of crossbills. *Ecological Monographs* 57: 251–67. https://doi.org/10.2307/2937083

Benkman, C.W. (1987b) Crossbill foraging behavior, bill structure, and patterns of food profitability. *Wilson Bulletin* 99: 351–68.

Benkman, C.W. (1988) Seed handling ability, bill structure, and the cost of bill specialization for crossbills. *Auk* 105: 715–19. https://doi.org/10.1093/auk/105.4.715

Benkman, C.W. (1989a) Intake rate maximization and the foraging behaviour of crossbills. *Ornis Scandinavica* 20: 65–8. https://doi.org/10.2307/3676710

Benkman, C.W. (1989b) On the foraging behavior of the Kauai Akepa. *'Elepaio* 49: 29–30.

Benkman, C.W. (1989c) On the evolution and ecology of island populations of crossbills. *Evolution* 43: 1324–30. https://doi.org/10.1111/j.1558-5646.1989.tb02581.x

Benkman, C.W. (1990) Foraging rates and the timing of crossbill reproduction. *Auk* 107: 376–86. https://doi.org/10.2307/4087622

Benkman, C.W. (1993a) Adaptation to single resources and the evolution of crossbill (*Loxia*) diversity. *Ecological Monographs* 63: 305–25. https://doi.org/10.2307/2937103

Benkman, C.W. (1993b) Logging, conifers, and the conservation of crossbills. *Conservation Biology* 7: 473–9. https://doi.org/10.1046/j.1523-1739.1993.07030473.x

Benkman, C.W. (1996) Are the ratios of bill crossing morphs in crossbills the result of frequency-dependent selection? *Evolutionary Ecology* 10: 119–26. https://doi.org/10.1007/BF01239352

Benkman, C.W. (1997) Feeding behavior, flock-size dynamics, and variation in sexual selection in crossbills. *Auk* 114: 163–78. https://doi.org/10.2307/4089158

Benkman, C.W. (1999) The selection mosaic and diversifying coevolution between crossbills and Lodgepole Pine. *American Naturalist* 154: S75–S91. https://doi.org/10.1086/303213

Benkman, C.W. (2003) Divergent selection drives the adaptive radiation of crossbills. *Evolution* 57: 1176–81. https://doi.org/10.1111/j.0014-3820.2003.tb00326.x

Benkman, C.W. (2013) Biotic interaction strength and the intensity of selection. *Ecology Letters* 16: 1054–560. https://doi.org/10.1111/ele.12138

Benkman, C.W. (2016) The natural history of the South Hills crossbill in relation to its impending extinction. *American Naturalist* 188: 589–601. https://doi.org/10.1086/688904

Benkman, C.W. (2017a) Matching habitat choice in nomadic crossbills appears most pronounced when food is most limiting. *Evolution* 71: 778–85. https://doi.org/10.1111/evo.13146

Benkman, C.W. (2017b) Crossbills were unlikely resident in the Bahamas; thus, there was no population to be extirpated. *Proceedings of the National Academy of Sciences USA* 114: E10031–2. https://doi.org/10.1073/pnas.1716928114

Benkman, C.W. (2020) White-winged Crossbill (*Loxia leucoptera*), version 1.0. In S.M. Billerman, ed., *Birds of the World*. Ithaca, NY: Cornell Lab of Ornithology. https://doi.org/10.2173/bow.whwcro.01

Benkman, C.W. (2024) The peculiar cousins: lessons from the ecology and evolution of crossbills. In G.C. Cardoso, R.J. Lopes and P.G. Mota, eds, *The Canary: Natural History, Science and Cultural Significance*, pp. 135–61. London: Academic Press. https://doi.org/10.1016/B978-0-443-15350-1.00005-8

Benkman, C.W. and Lindholm, A.K. (1991) The advantages and evolution of a morphological novelty. *Nature* 349: 519–20. https://doi.org/10.1038/349519a0

Benkman, C.W. and Mezquida, E.T. (2015) Phenotypic selection exerted by a seed predator is replicated in space, time, and among prey species. *American Naturalist* 186: 682–91. https://doi.org/10.1086/683131

Benkman, C.W. and Miller, R.E. (1996) Morphological evolution in response to fluctuating selection. *Evolution* 50: 2499–504. https://doi.org/10.2307/2410716

Benkman, C.W. and Parchman, T.P. (2009) Coevolution between crossbills and Black Pine: the importance of competitors, forest area, and resource stability. *Journal of Evolutionary Biology* 22: 942–53. https://doi.org/10.1111/j.1420-9101.2009.01703.x

Benkman, C.W. and Parchman, T.L. (2013) When directional selection reduces geographic variation in traits mediating species interactions. *Ecology and Evolution* 3: 961–70. https://doi.org/10.1002/ece3.518

Benkman, C.W. and Pulliam, H.R. (1988) The comparative feeding rates of North American sparrows and finches. *Ecology* 69: 1195–9. https://doi.org/10.2307/1941274

Benkman, C.W. and Siepielski, A.M. (2004) A keystone selective agent? Pine squirrels and the frequency of serotiny in Lodgepole Pine. *Ecology* 85: 2082–7. https://doi.org/10.1890/04-0177

Benkman, C.W. and Young, M. (2020) Red Crossbill (Loxia curvirostra). In P.G. Rodewald, ed., *The Birds of North America Online*. Ithaca, NY: Cornell Lab of Ornithology. https://doi.org/10.2173/bow.redcro.01

Benkman, C.W., Balda, R.P. and Smith, C.C. (1984) Adaptations for seed dispersal and the compromises due to seed predation in limber pine. *Ecology* 65: 632–42. https://doi.org/10.2307/1941426

Benkman, C.W., Holimon, W.C. and Smith, J.W. (2001) The influence of a competitor on the geographic mosaic of coevolution between crossbills and Lodgepole Pine. *Evolution* 55: 282–94. https://doi.org/10.1111/j.0014-3820.2001.tb01293.x

Benkman, C.W., Parchman, T.L., Favis, A. and Siepielski, A.M. (2003) Reciprocal selection causes a coevolutionary arms race between crossbills and Lodgepole Pine. *American Naturalist* 162: 182–94. https://doi.org/10.1086/376580

Benkman, C.W., Siepielski, A.M. and Parchman, T.L. (2008) The local introduction of strongly interacting species and the loss of geographic variation in species and species interactions. *Molecular Ecology* 17: 395–404. https://doi.org/10.1111/j.1365-294X.2007.03368.x

Benkman, C.W., Smith, J.W., Keenan, P.C., Parchman, T.L. and Santisteban, L. (2009). A new species of Red Crossbill (Fringillidae: *Loxia*) from Idaho. *Condor* 111: 169–76. https://doi.org/10.1525/cond.2009.080042

Benkman, C.W., Parchman, T.L. and Mezquida, E.T. (2010) Patterns of coevolution in the adaptive radiation of crossbills. Annals of the New York Academy of Sciences 1206: 1–16. https://doi.org/10.1111/j.1749-6632.2010.05702.x

Benkman, C.W., Fetz, T. and Talluto, M.V. (2012) Variable resource availability when resource replenishment is constant: the coupling of predators and prey. *Auk* 129: 115–23. https://doi.org/10.1525/auk.2011.11069

Benkman, C.W., Smith, J.W., Maier, M., Hansen, L. and Talluto, M.V. (2013) Consistency and variation in phenotypic selection exerted by a community of seed predators. *Evolution* 67: 157–69. https://doi.org/10.1111/j.1558-5646.2012.01736.x

Benkman, C.W., Brock, C.D., Parchman, T.L. and Porter, C.K. (2022) Response to Hill and Powers: It is irrelevant that the mode and tempo of Cassia Crossbill speciation is not typical for birds. *Journal of Avian Biology* 2022: e02967. https://doi.org/10.1111/jav.02967

Bentz, B.J., Régnière, J., Fettig, C.J., Hansen, E.M., Hayes, J.L., Hicke, J.A., Kelsey, R.G., Negrón, J.F. and Seybold, S.J. (2010) Climate change and bark beetles of the western United States and Canada: direct and indirect effects. *BioScience* 60: 602–13. https://doi.org/10.1525/bio.2010.60.8.6

Betts, M.G., Yang, Z., Hadley, A.S., Smith, A.C., Rousseau, J.S., Northrup, J.M., Nocera, J.J., Gorelick, N. and Gerber, B.D. (2022) Forest degradation drives widespread avian habitat and population declines. *Nature Ecology and Evolution* 6: 709–19. https://doi.org/10.1038/s41559-022-01737-8

Birkhead, T.R. and Montgomerie, R. (2020) Three decades of sperm competition in birds. *Philosophical Transactions of the Royal Society of London B* 375: 20200208-7. https://doi.org/10.1098/rstb.2020.0208

Boag, P.T. and van Noordwijk, A. (1987) Quantitative genetics. In F. Cooke. and P.A. Buckley, eds, *Avian Genetics: A Population and Ecological Approach*, pp. 43–78. New York: Academic Press. https://doi.org/10.1016/B978-0-12-187571-8.50009-5

Bock, C.E. and Lepthien, L.W. (1976) Synchronous eruptions of boreal seed-eating birds. *American Naturalist* 110: 559–71. https://doi.org/10.1086/283091

Boehm, M.M.A., Guevara-Apaza, D., Jankowski, J.E. and Cronk, Q.C.B. (2022) Floral phenology of an Andean bellflower and pollination by Buff-tailed Sicklebill Hummingbird. *Ecology and Evolution* 12: e8988. https://doi.org/10.1002/ece3.8988

Bolger, D.T., Newmark, W.D., Morrison, T.A. and Doak, D.F. (2007) The need for integrative approaches to understand and conserve migratory ungulates. *Ecology Letters* 11: 63–77. https://doi.org/10.1111/j.1461-0248.2007.01109.x

Bolnick, D.I. and Otto, S.P. (2013) The magnitude of local adaptation under genotype-dependent dispersal. *Ecology and Evolution* 3: 4722–35. https://doi.org/10.1002/ece3.850

Bousquet, C.A.H., Sumpter, D.J.T. and Manser, M.B. (2011) Moving calls: a vocal mechanism underlying quorum decisions in cohesive groups. *Proceedings of the Royal Society of London B* 278: 1482–8. https://doi.org/10.1098/rspb.2010.1739

Bowman, R.I. (1961) Morphological differentiation and adaptation in the Galápagos finches. *University of California Publications in Zoology* 58: 1–302.

Brink, C.H. and Dean, F.C. (1966) Spruce seed as a food of red squirrels and flying squirrels in interior Alaska. *Journal of Wildlife Management* 30: 503–12. https://doi.org/10.2307/3798741

Brouwer, L. and Griffith, S.C. (2019) Extra-pair paternity in birds. *Molecular Ecology* 28: 4864–82. https://doi.org/10.1111/mec.15259

Brown, C.D. and Johnstone, J.F. (2012) Once burned, twice shy: repeat fires reduce seed availability and alter substrate constraints on *Picea mariana* regeneration. *Forest Ecology and Management* 266: 34–41. https://doi.org/10.1016/j.foreco.2011.11.006

Brown, J.H. (1995) *Macroecology*. Chicago: University of Chicago Press.

Brown Jr., W.L. (1957) Centrifugal speciation. *Quarterly Review of Biology* 32: 247–77. https://doi.org/10.1086/401875

Burns, R.M. and Honkala, B.H. (1990) *Silvics of North America*, Vol. 1. United States Department of Agriculture, Agricultural Handbook 654.

Cadieux, P., Boulanger, Y., Cyr, D., Taylor, A.R., Price, D.T., Sólymos, P., Stralberg, D., Chen, H.Y.H., Brecka, A. and Tremblay, J.A. (2020) Projected effects of climate change on boreal bird community accentuated by anthropogenic disturbances in western boreal forest, Canada. *Diversity and Distributions* 26: 668–82. https://doi.org/10.1111/ddi.13057

Cahill, A.E., Aiello-Lammens, M.E., Fisher-Reid, M.C., Hua, X., Karanewsky, C.J., Ryu, H.Y., Sbeglia, G.C., Spagnolo, F., Waldron, J.B., Warsi, O. and Wiens, J.J. (2013) How does climate change cause extinction? *Proceedings of the Royal Society of London B* 280: 20121890. https://doi.org/10.1098/rspb.2012.1890

Ceballos, G., Ehrlich, P.R. and Dirzo, R. (2017) Biological annihilation via the ongoing sixth mass extinction signaled by vertebrate population losses and declines. *Proceedings of the National Academy of Sciences USA* 114: E6089–96. https://doi.org/10.1073/pnas.1704949114

Chai, T.S. and Hansen, H.L. (1952) Characteristics of Black Spruce seed from cones of different ages. Minnesota Forestry Notes 2, University of Minnesota, St. Paul, MN.

Charnov, E.L. (1976) Optimal foraging: the marginal value theorem. *Theoretical Population Biology* 9: 129–36. https://doi.org/10.1016/0040-5809(76)90040-X

Chesser, R.T., Burns, K.J., Cicero, C., Dunn, J.L., Kratter, A.W., Lovette, I.J., Rasmussen, P.C., Remsen, J.V., Rising, J.D., Stotz, D.F. and Winker, K. (2017) Fifty-eighth supplement to the American Ornithological Society's *Check-list of North American Birds*. *Auk: Ornithological Advances* 134: 751–73. https://doi.org/10.1642/AUK-17-72.1

Chesser, R.T., Billerman, S.M., Burns, K.J., Cicero, C., Dunn, J.L., Hernández-Baños, B.E., Jiménez, R.A., Kratter, A.W., Mason, N.A., Rasmussen, P.C., Remsen, J.V., Stotz, D.F. and Winker, K. (2022) Sixty-third supplement to the American Ornithological Society's *Check-list of North American Birds*. *Ornithology* 139: ukac020. https://doi.org/10.1093/ornithology/ukac020

Christidis, N., Jones, G.S. and Stott, P.A. (2015) Dramatically increasing chance of extremely hot summers since the 2003 European heatwave. *Nature Climate Change* 5: 46–50. https://doi.org/10.1038/nclimate2468

Clark, C.W. and Mangel, M. (1984) Foraging and flocking strategies: Information in an uncertain environment. *American Naturalist* 123: 626–41. https://doi.org/10.1086/284228

Clark, C.W. and Mangel, M. (1986) The evolutionary advantages of group foraging. *Theoretical Population Biology* 30: 45–75. https://doi.org/10.1016/0040-5809(86)90024-9

Clark, P.U., Shakun, J.D., Rosenthal, Y., Köhler, P. and Bartlein, P. (2024) Global and regional temperature change over the past 4.5 million years. *Science* 383: 884–90. https://doi.org/10.1126/science.adi1908

Clayton, D.H. and Walther, B.A. (2001) Influence of host ecology and morphology on the diversity of Neotropical bird lice. *Oikos* 94: 455–67. https://doi.org/10.1034/j.1600-0706.2001.940308.x

Clayton, D.H., Moyer, B.R., Bush, S.E., Jones, T.G., Gardiner, D.W., Rhodes, B.B. and Goller, F. (2005) Adaptive significance of avian beak morphology for ectoparasite control. *Proceedings of the Royal Society of London B* 272: 811–17. https://doi.org/10.1098/rspb.2004.3036

Clayton, D.H., Koop, J.A.H., Harbison, C.W., Moyer, B.R. and Bush, S.E. (2010) How birds combat ectoparasites. *Open Ornithology Journal* 3: 41–71. https://doi.org/10.2174/1874453201003010041

Clayton, D.H., Bush, S.E. and Johnson, K.P. (2016) *Coevolution of Life on Hosts: Integrating Ecology and History*. Chicago: University of Chicago Press. https://doi.org/10.7208/chicago/9780226302300.001.0001

Clouet, M. (2000) The breeding biology of the Common Crossbill *Loxia curvirostra* in the central Pyrenees. *Bird Study* 47: 186–94. https://doi.org/10.1080/00063650009461174

Clouet, M. (2003) Taille du bec et période de reproduction chez les beccroisés des forêts de pins. *Revue d'Ecologie (La Terre et al Vie)* 58: 419–33. https://doi.org/10.3406/revec.2003.5342

Clouet, M. and Goar, J. (1999) Le Beccroisés d'Annam *Loxia curvirostra meridionalis*. *Alauda* 67: 53–62.

Clouet, M. and Goar, J. (2001) Note on the Philippine Crossbill *Loxia curvirostra luzoniensis*. *Alauda* 69: 331–4.

Coffey, K., Benkman, C.W. and Milligan, B.G. (1999) The adaptive significance of spines on pine cones. *Ecology* 80: 1221–9. https://doi.org/10.1890/0012-9658(1999)080[1221:TASOSO]2.0.CO;2

Conradt, L. and Roper, T.J. (2005) Consensus decision making in animals. *Trends in Ecology and Evolution* 20: 449–56. https://doi.org/10.1016/j.tree.2005.05.008

Coop, J.D., Parks, S.A., Stevens-Rumann, C.S., Crausbay, S.D., Higuera, P.E., Hurteau, M.D., Tepley, A., Whitman, E., Assal, T., Collins, B.M., Davis, K.T., Dobrowski, S., Falk, D.A., Fornwalt, P.J., Fulé, P.Z., Harvey, B.J., Kane, V.R., Littlefield, C.E., Margolis, E.Q., North, M., Parisien, M.-A., Prichard, S. and Rodman, K.C. (2020) Wildfire-driven forest conversion in western North American landscapes. *BioScience* 70: 659–73. https://doi.org/10.1093/biosci/biaa061

Coops, N.C. and Waring, R.H. (2011) A process-based approach to estimate Lodgepole Pine (*Pinus contorta* Dougl.) distribution in the Pacific Northwest under climate change. *Climatic Change* 105: 313–28. https://doi.org/10.1007/s10584-010-9861-2

Cornelius, J.M. and Hahn, T.P. (2012) Seasonal pre-migratory fattening and increased activity in a nomadic and irruptive migrant, the Red Crossbill *Loxia curvirostra*. *Ibis* 154: 693–702. https://doi.org/10.1111/j.1474-919X.2012.01266.x

Coyne, J.A. and Orr, H.A. (1989) Patterns of speciation in *Drosophila*. *Evolution* 43: 362–81. https://doi.org/10.2307/2409213; https://doi.org/10.1111/j.1558-5646.1989.tb04233.x

Coyne, J.A. and Orr, H.A. (2004) *Speciation*. Sunderland: Sinauer.

Craig, T.P., Itami, J.K. and Horner, J.D. (2007) Geographic variation in the evolution and coevolution of a tritrophic interaction. *Evolution*. 61: 1137–52. https://doi.org/10.1111/j.1558-5646.2007.00099.x

Cramp, S. and Perrins, C.M. (1994) *The Birds of the Western Palearctic*, Vol. 8: *Crows to Finches*. Oxford: Oxford University Press.

Critchfield, W.B. (1980) *Genetics of Lodgepole Pine*. United States Forest Service Research Paper WO-37.

Critchfield, W.B. (1985) The late Quaternary history of Lodgepole and Jack Pines. *Canadian Journal of Forest Research* 15: 749–72. https://doi.org/10.1139/x85-126

Critchfield, W.B. and Little Jr., E.L. (1966) *Geographic Distribution of the Pines of the World.* Miscellaneous Publications 991. United States Department of Agriculture, Washington, D. C. https://doi.org/10.5962/bhl.title.66393

Darwin, C. (1859) *On the Origin of Species by Means of Natural Selection.* London: John Murray.

Daubenmire, R. (1968) Some geographic variations in *Picea sitchensis* and their ecologic interpretation. *Canadian Journal of Botany* 46: 787–98. https://doi.org/10.1139/b68-110

Davis, K., Dobrowski, S., Higuera, P., Holden, Z., Veblen, T., Rother, M., Parks, S., Sala, A. and Maneta, M. (2019) Wildfires and climate change push low-elevation forests across a critical climate threshold for tree regeneration. *Proceedings of the National Academy of Sciences USA* 116: 6193–8. https://doi.org/10.1073/pnas.1815107116

Davis, K.T., Robles, M.D., Kemp, K.B., Higuera, P.E., Chapman, T., Metlen, K.L., Peeler, J.L., Rodman, K.C., Woolley, T., Addington, R.N., Buma, B.J., Cansler, C.A., Case, M.J., Collins, B.M., Coop, J.D., Dobrowski, S.Z., Gill, N.S., Haffey, C., Harris, L.B., Harvey, B.J., Haugo, R.D., Hurteau, M.D., Kulakowski, D., Littlefield, C.E., McCauley, L.A., Povak, N., Shive, K.L., Smith, E., Stevens, J.T., Stevens-Rumann, C.S., Taylor, A.H., Tepley, A.J., Young, D.J.N., Andrus, R.A., Battaglia, M.A., Berkey, J.K., Busby, S.U., Carlson, A.R., Chambers, M.E., Dodson, E.K., Donato, D.C., Downing, W.M., Fornwalt, P.J., Halofsky, J.S., Hoffman, A., Holz, A., Iniguez, J.M., Krawchuk, M.A., Kreider, M.R., Larson, A.J., Meigs, G.W., Roccaforte, J.P., Rother, M.T., Safford, H., Schaedel, M., Sibold, J.S., Singleton, M.P., Turner, M.G., Urza, A.K., Clark-Wolf, K.D., Yocom, L., Fontaine, J.B. and Campbell, J.L. (2023) Reduced fire severity offers near-term buffer to climate-driven declines in conifer resilience across the western United States. *Proceedings of the National Academy of Sciences USA* 120: e2208120120. https://doi.org/10.1073/pnas.2208120120

Davis, O.K., Sheppard, J.C. and Robertson, S. (1986) Contrasting climatic histories for the Snake River Plain, Idaho, resulting from multiple thermal maxima. *Quaternary Research* 26: 321–39. https://doi.org/10.1016/0033-5894(86)90093-1

De Meester, L., Gómez, A., Okamura, B. and Schwenk, K. (2002) The monopolization hypothesis and the dispersal–gene flow paradox in aquatic organisms. *Acta Oecologica* 23: 121–135. https://doi.org/10.1016/S1146-609X(02)01145-1

Dibnah, A.J., Herbert-Read, J.E., Boogert, N.J., McIvor, G.E., Jolles, J.W. and Thornton, A. (2022) Vocally mediated consensus decisions govern mass departures from Jackdaw roosts. *Current Biology* 32: R455–6. https://doi.org/10.1016/j.cub.2022.04.032

Dochtermann, N.A., Schwab, T., Berdal, M.A., Dalos, J. and Royauté, R. (2019) The heritability of behavior: a meta-analysis. *Journal of Heredity* 110: 403–10. https://doi.org/10.1093/jhered/esz023

Doligez, B., Danchin, E. and Clobert, J. (2002) Public information and breeding habitat selection in a wild bird population. *Science* 297: 1168–70. https://doi.org/10.1126/science.1072838

Duffy, P.B. and Tebaldi, C. (2012) Increasing prevalence of extreme summer temperatures in the U.S. *Climatic Change* 111: 487–95. https://doi.org/10.1007/s10584-012-0396-6

Dunn, E.H. (2019) Dynamics and population consequences of irruption in the Red-breasted Nuthatch (*Sitta canadensis*). *Auk* 136: 1–10.

Ebenhard, T. (1988) Introduced birds and mammals and their ecological effects. *Swedish Wildlife Research* 13: 1–107.

Edelaar, P. (2008) Rediscovery of a second kind of crossbill for the Himalayan region, and the hypothesis that ecological opportunity drives crossbill diversification. *Ibis* 150: 405–8. https://doi.org/10.1111/j.1474-919X.2007.00766.x

Edelaar, P. (2022) Sexual selection may not often reduce gene flow between locally adapted populations. A review of some evidence, and suggestions for better tests. *Frontiers in Ecology and Evolution* 10: 804910. https://doi.org/10.3389/fevo.2022.804910

Edelaar, P. and Terpstra, K. (2004) Is the nominate subspecies of the Common Crossbill *Loxia c. curvirostra* polytypic? I. Morphological differences among years at a single site. *Ardea* 92: 93–102.

Edelaar, P. and Benkman, C.W. (2006) Replicated population divergence caused by localized coevolution? A test of three hypotheses in the Red Crossbill–Lodgepole Pine system. *Journal of Evolutionary Biology* 19: 1651–9. https://doi.org/10.1111/j.1420-9101.2006.01113.x

Edelaar, P., Postma, E., Knops, P. and Phillips, R. (2005) No support for a genetic basis of mandible crossing direction in crossbills (*Loxia* spp). *Auk* 122: 1123–39. https://doi.org/10.1093/auk/122.4.1123

Edelaar, P., Siepielski, A.M. and Clobert, J. (2008) Matching habitat choice causes directed gene flow: a neglected dimension in evolution and ecology. *Evolution* 62: 2462–72. https://doi.org/10.1111/j.1558-5646.2008.00459.x

Edelaar, P., Alonso, D., Lagerveld, S., Senar, J.C. and Björklund, M. (2012) Population differentiation and restricted gene flow in Spanish crossbills: not isolation-by-distance but isolation-by-ecology. *Journal of Evolutionary Biology* 25: 417–30. https://doi.org/10.1111/j.1420-9101.2011.02443.x

Ehrlich, P.R. and Raven, P.H. (1964) Butterflies and plants: a study in coevolution. *Evolution* 18: 586–608. https://doi.org/10.2307/2406212

Elliott, P.F. (1974) Evolutionary responses of plants to seed-eaters: pine squirrel predation on Lodgepole Pine. *Evolution* 28: 221–31. https://doi.org/10.2307/2407323

Elliott, P.F. (1988a) Foraging behavior of a central-place forager: field tests of theoretical predictions. *American Naturalist* 131: 159–74. https://doi.org/10.1086/284784

Elliott, P.F. (1988b) The influence of population density and body size on the behavioral ecology of the pine squirrel (*Tamiasciurus hudsonicus*). Natural History Paper no. 1. Mountain Research Center, University of Colorado, Boulder, CO.

Enright, N.J., Marsula, R., Lamont, B.B. and Wissel, C. (1998) The ecological significance of canopy seed storage in fire-prone environments: a model for non-sprouting shrubs. *Journal of Ecology* 86: 946–59. https://doi.org/10.1046/j.1365-2745.1998.00312.x

Enright, N.J., Fontaine, J.B., Bowman, D.M., Bradstock, R.A. and Williams, R.J. (2015) Interval squeeze: altered fire regimes and demographic responses interact to threaten woody species persistence as climate changes. *Frontiers in Ecology and the Environment* 13: 265–72. https://doi.org/10.1890/140231

Feinsinger, P. (1983) Coevolution and pollination. In D.J. Futuyma and M. Slatkin, eds, *Coevolution*, pp. 282–310. Sunderland: Sinauer.

Flannery, T. (1994) *The Future Eaters: An Ecological History of the Australasian Lands and People*. Sydney: Reed Books.

Fletcher, Q.E., Boutin, S., Lane, J.E., LaMontagne, J.M., McAdam, A.G., Krebs, C.J. and Humphries, M.M. (2010) The functional response of a hoarding seed predator to mast seeding. *Ecology* 91: 2673–83. https://doi.org/10.1890/09-1816.1

Forde, S.E., Thompson, J.N. and Bohannan, B.J.M. (2004) Adaptation varies through space and time in a coevolving host–parasitoid interaction. *Nature* 431: 841–4. https://doi.org/10.1038/nature02906

Forister, M.L., Novotny, V., Panorska, A.K., Baje, L., Basset, Y., Butterill, P.T., Cizek, L., Coley, P.D., Dem, F., Diniz, I.R., Drozd, P., Fox, M., Glassmire, A.E., Hazen, R., Hrcek, J., Jahner, J.P., Kaman, O., Kozubowski, T.J., Kursar, T.A., Lewis, O.T., Lill, J., Marquis, R.J., Miller, S.E., Morais, H.C., Murakami, M., Nickel, H., Pardikes, N.A., Ricklefs, R.E., Singer, M.S., Smilanich, A.M., Stireman, J.O., Villamarín-Cortez, S., Vodka, S., Volf, M., Wagner, D.L., Walla, T., Weiblen, G.D. and Dyer, L.A. (2015) The global distribution of diet breadth in insect herbivores. *Proceedings of the National Academy of Sciences USA* 112: 442–7. https://doi.org/10.1073/pnas.1423042112

Fowells, H.A. (1965) *Silvics of Forest Trees of the United States*. Agricultural Handbook Number 271, United States Department of Agriculture, Washington, DC.

Freed, L.A., Medeiros, M.C. and Bodner, G.R. (2008) Explosive increase in ectoparasites in Hawaiian forest birds. *Journal of Parasitology* 94: 1009–21. https://doi.org/10.1645/GE-1378.1

Freeman, B.G., Weeks, T., Schluter, D. and Tobias, J. (2022) The latitudinal gradient in rates of evolution for bird beaks, a species interaction trait. *Ecology Letters* 25: 635–46. https://doi.org/10.1111/ele.13726

Friedman, N.R., Miller, E.T., Ball, J.R., Kasuga, H., Remeš, V. and Economo, E.P. (2019) Evolution of a multifunctional trait: shared effects of foraging ecology and thermoregulation on beak morphology, with consequences for song evolution. *Proceedings of the Royal Society of London B* 286: 20192474-10. https://doi.org/10.1098/rspb.2019.2474

Futuyma, D.J. (2013) *Evolution*, 3rd edn. Sunderland: Sinauer.

Gao, J., Tomlinson, K.W., Zhao, W., Wang, B., Lapuz, R.S., Liu, J., Pasion, B.O., Hai, B.T., Chanthayod, S., Chen, J. and Wang, X. (2024) Phylogeography and introgression between *Pinus kesiya* and *Pinus yunnanensis* in Southeast Asia. *Journal of Systematics and Evolution* 62: 120–134. https://doi.org/10.1111/jse.12949

Germain, R.R., Feng, S., Chen, G., Graves, G.R., Tobias, J.A., Rahbek, C., Lei, F., Fjeldså, J., Hosner, P.A., Gilbert, M.T.P., Zhang, G. and Nogués-Bravo, D. (2023) Species-specific traits mediate avian demographic responses under past climate change. *Nature Ecology and Evolution* 7: 862–72. https://doi.org/10.1038/s41559-023-02055-3

Gil, M.A., Emberts, Z., Jones, H. and St. Mary, C.M. (2017) Social information on fear and food drives animal grouping and fitness. *American Naturalist* 189: 227–41. https://doi.org/10.1086/690055

Gill, F.B. and Wolf, L.L. (1977) Nonrandom foraging by sunbirds in a patchy environment. *Ecology* 58: 1284–96. https://doi.org/10.2307/1935081

Godfrey, W.E. (1950) *Birds of the Cypress Hills and Flotten Lakes Regions, Saskatchewan*. National Museum of Canada, Department of Resources and Development, Bulletin Number 120.

Godfrey, W.E. (1986) *The Birds of Canada*. National Museum of Natural Sciences, National Museums of Canada, Ottawa, Canada.

Gómez-Blanco, D., Santoro, S., Borrás, A., Cabrera, J., Senar, J.C. and Edelaar, P. (2019) Beak morphology predicts apparent survival of crossbills: due to selective survival or selective dispersal? *Journal of Avian Biology* 50: e02107. https://doi.org/10.1111/jav.02107

Gómez-Llano, M., Bassar, R.D., Svensson, E.I., Tye, S.P. and Siepielski, A.M. (2024) Meta-analytical evidence for frequency-dependent selection across the tree of life. *Ecology Letters* 27: e14477. https://doi.org/10.1111/ele.14477

Gomulkiewicz, R., Drown, D.M., Dybdahl, M.F., Godsoe, W., Nuismer, S.L., Pepin, K.M., Ridenhour, B.J., Smith, C.I. and Yoder, J.B. (2007) Dos and don'ts of testing the geographic mosaic theory of coevolution. *Heredity* 98: 249–58. https://doi.org/10.1038/sj.hdy.6800949

Grant, B.R. and Grant, P.R. (1989) *Evolutionary Dynamics of a Natural Population: The Large Cactus Finch of the Galápagos*. Chicago: University of Chicago Press.

Grant, B.R. and Grant, P.R. (2010) Songs of Darwin's finches diverge when a new species enters the community. *Proceedings of the National Academy of Sciences USA* 107: 20156–63. https://doi.org/10.1073/pnas.1015115107

Grant, P.R. (1986) *Ecology and Evolution of Darwin's Finches*. Princeton: Princeton University Press.

Grant, P.R. and Grant, B.R. (2008) *How and Why Species Multiply*. Princeton: Princeton University Press. https://doi.org/10.1515/9781400837946

Grant, P.R. and Grant, B.R. (2014) *40 Years of Evolution: Darwin's Finches on Daphne Major Island*. Princeton: Princeton University Press. https: //doi.org/10.1515/9781400851300

Greenberg, R., Cadena, V., Danner, R.M. and Tattersall, G. (2012) Heat loss may explain bill size differences between birds occupying different habitats. *PLoS ONE* 7: e40933. https://doi.org/10.1371/journal.pone.0040933

Greene, D.F., Kane, J.M., Pounden, E. and Michaletz, S.T. (2024) Cone allometry and seed protection from fire are similar in serotinous and nonserotinous conifers. *New Phytologist* 242: 93–106. https://doi.org/10.1111/nph.19578

Grinnell, J. (1931) Some angles in the problem of bird migration. *Auk* 48: 22–32. https://doi.org/10.2307/4076942

Griscom, L. (1937) A monographic study of the Red Crossbill. *Proceedings of the Boston Society of Natural History* 41: 77–210.

Groth, J.G. (1988) Resolution of cryptic species in Appalachian Red Crossbills. *Condor* 90: 745–60. https://doi.org/10.2307/1368832

Groth, J.G. (1993a) Evolutionary differentiation in morphology, vocalizations, and allozymes among nomadic sibling species in the North American Red Crossbill (*Loxia curvirostra*) complex. *University of California Publications in Zoology* 127: 1–143.

Groth, J.G. (1993b) Call matching and positive assortative mating in Red Crossbills. *Auk* 110: 398–401.

Gurnell, J. (1984) Home range, territoriality, caching behaviour and food supply of the Red Squirrel (*Tamiasciurus hudsonicus fremonti*) in a subalpine Lodgepole Pine forest. *Animal Behaviour* 32: 1119–31. https://doi.org/10.1016/S0003-3472(84)80228-6

Haapanen, A. (1966) Bird fauna of the Finnish forests in relation to forest succession. *Annales Zoologici Fennici* 3: 176–200.

Hahn, T.P. (1998) Reproductive seasonality in an opportunistic breeder, the Red Crossbill, *Loxia curvirostra*. *Ecology* 79: 2365–75. https://doi.org/10.2307/176828

Hannah, K.C., Kardynal, K.J. and Hobson, K.A. (2020) Birds of a feather don't always flock together: variation in molt origins and movement patterns of winter finches in Ontario. *Journal of Ornithology* 161: 609–20. https://doi.org/10.1007/s10336-020-01760-1

Hannerz, M., Almqvist, C. and Hornfeldt, R. (2002) Timing of seed dispersal in *Pinus sylvestris* stands in central Sweden. *Silva Fennica* 36: 757–65. https://doi.org/10.14214/sf.518

Hansen, W.D. and Turner, M.G. (2019) Origins of abrupt change? Postfire subalpine conifer regeneration declines nonlinearly with warming and drying. *Ecological Monographs* 89: e01340–21. https://doi.org/10.1002/ecm.1340

Hembry, D.H., Yoder, J.B. and Goodman, K.R. (2014) Coevolution and the diversification of life. *American Naturalist* 184: 425–38. https://doi.org/10.1086/677928

Hendry, A.P., Day, T. and Taylor, E.B. (2001) Population mixing and the adaptive divergence of quantitative traits in discrete populations: a theoretical framework for empirical tests. *Evolution* 55: 459–66. https://doi.org/10.1111/j.0014-3820.2001.tb00780.x

Hicke, J.A., Xu, B., Meddens, A.J.H. and Egan J.M. (2020) Characterizing recent bark beetle-caused tree mortality in the western United States from aerial surveys. *Forest Ecology and Management* 475: 118402. https://doi.org/10.1016/j.foreco.2020.118402

Hill, G.E., and Powers, M.J. (2021) Point of view: Ecomorphs are not species: the case of locally adapted populations of Red Crossbills. *Journal of Avian Biology* 52: e02896. https://doi.org/10.1111/jav.02896

Hori, M. (1993) Frequency-dependent natural selection in the handedness of scale-eating cichlid fish. *Science* 260: 216–19. https://doi.org/10.1126/science.260.5105.216

Howell, T. (1972) Birds of the lowland pine savanna of northeastern Nicaragua. *Condor* 74: 316–40. https://doi.org/10.2307/1366592

Hudson, R.R. and Coyne, J. (2002) Mathematical consequences of the genealogical species concept. *Evolution* 56: 1557–65. https://doi.org/10.1111/j.0014-3820.2002.tb01467.x

Hutchinson, G.E. (1959) Homage to Santa Rosalia or why are there so many kinds of animals? *American Naturalist* 93: 145–59. https://doi.org/10.1086/282070

Hutchinson, M.C., Dobson, A.P. and Pringle, R.M. (2022) Dietary abundance distributions: dominance and diversity in vertebrate diets. *Ecology Letters* 25: 992–1008. https://doi.org/10.1111/ele.13948

Hutto, R.L. (1988) Foraging behavior patterns suggest a possible cost associated with participation in mixed-species bird flocks. *Oikos* 51: 79–83. https://doi.org/10.2307/3565809

Irwin, K. (2010) A new and cryptic call type of the Red Crossbill. *Western Birds* 41: 10–25.

Jackson, S.T., Overpeck, J.T., Webb-III, T., Keattch, S.E. and Anderson, K.H. (1997) Mapped plant-macrofossil and pollen records of late Quaternary vegetation change in eastern North America. *Quaternary Science Reviews* 16: 1–70. https://doi.org/10.1016/S0277-3791(96)00047-9

Jaime, L., Batllori, E. and Lloret, F. (2024) Bark beetle outbreaks in coniferous forests: a review of climate change effects. *European Journal of Forest Research* 143: 1–17. https://doi.org/10.1007/s10342-023-01623-3

Jansson, R. and Dynesius, M. (2002) The fate of clades in a world of recurrent climate change: Milankovitch oscillations and evolution. *Annual Reviews of Ecology and Systematics* 33: 741–77. https://doi.org/10.1146/annurev.ecolsys.33.010802.150520

Janzen, D.H. (1969) Seed-eaters versus seed size, number, toxicity and dispersal. *Evolution* 23: 1–27. https://doi.org/10.2307/2406478

Jeschke, J.M. (2008) Across islands and continents, mammals are more successful invaders than birds. *Diversity and Distributions* 14: 913–16. https://doi.org/10.1111/j.1472-4642.2008.00488.x

Jin, W.-T., Gernandt, D.S., Wehenkei, C., Xia, X.-M. and Wei, X.-X. (2021) Phylogenomic and ecological analyses reveal the spatiotemporal evolution of global pines. *Proceedings of the National Academy of Sciences USA* 118: e2022302118. https://doi.org/10.1073/pnas.2022302118

Johnstone, J.F. and Chapin III, F.S. (2006) Fire interval effects on successional trajectory in boreal forests of northwest Canada. *Ecosystems* 9: 268–77. https://doi.org/10.1007/s10021-005-0061-2

Joller, S., Bertschinger, F., Kump, E., Spin, A., von Rotz, A., Scheizer-Gorgas, D., Drögemüller, C. and Flury, C. (2018) Crossed beaks in a local Swiss chicken breed. *BMC Veterinary Research* 14: 68. https://doi.org/10.1186/s12917-018-1398-z

Joseph, L. (1986) Seed-eating birds of southern Australia. In H.A. Ford and D.C. Paton, eds, *The Dynamic Partnership: Birds and Plants in Southern Australia*, pp. 85–93. Handbook of the Flora and Fauna of South Australia Series. South Australian Government, South Australia.

Juutinen, P. (1953) Uber nahrung und forstwirtschaftliche bedeutung des fichtenkrenzschnabels (*L. curvirostra*). *Comment. Inst. Forestalis Fenniae* 41: 1–41.

Kalka, M.B., Smith, A.R. and Kalko, E.K.V. (2008) Bats limit arthropods and herbivory in a tropical forest. *Science* 320: 71. https://doi.org/10.1126/science.1153352

Kaltz, O., Gandon, S., Michalakis, Y. and Shykoff, J.A. (1999) Local maladaptation in the anther-smut fungus *Microbotryum violaceum* to its host plant *Silene latifolia*: evidence from a cross-inoculation experiment. *Evolution* 53: 395–407. https://doi.org/10.2307/2640776

Kaya, Z. and Temerit, A. (1994) Genetic structure of marginally located *Pinus nigra* var *pallasiana* populations in central Turkey. *Silvae Genetica* 43: 272–6.

Keeley, J.E., Ne'eman, G. and Fotheringham, C.J. (1999) Immaturity risk in a fire-dependent pine. *Journal of Mediterranean Ecology* 1: 41–7.

Keen, F.P. (1958) *Cone and Seed Insects of Western Forest Trees*. Technical Bulletin 1169, United States Department of Agriculture, Washington, D.C.

Keenan, P.C. and Benkman, C.W. (2008) Call imitation and call modification in Red Crossbills. *Condor* 110: 93–101. https://doi.org/10.1525/cond.2008.110.1.93

Kelly, D. and Sork, V.L. (2002) Mast seeding in perennial plants: why, how, where? *Annual Review of Ecology and Systematics* 33: 427–47. https://doi.org/10.1146/annurev.ecolsys.33.020602.095433

Kenward, R.E. (1978) Hawks and doves: attack success and selection in goshawk flights at woodpigeons. *Journal of Animal Ecology* 47: 449–60. https://doi.org/10.2307/3793

Khalil, M.A.K. (1984) Genetics of cone morphology of Black Spruce (*Picea mariana* Mill, B.S.P.) in Newfoundland, Canada. *Silvae Genetica* 33: 101–9.

Kimmins, J.P. (2003) Old-growth forest: an ancient and stable sylvan equilibrium, or a relatively transitory ecosystem condition that offers people a visual and emotional feast? Answer—it depends. *Forestry Chronicle* 79: 429–40. https://doi.org/10.5558/tfc79429-3

Kleven, O., Bjerke, B.A. and Lifjeld, J.T. (2008) Genetic monogamy in the Common Crossbill (*Loxia curvirostra*). *Journal of Ornithology* 149: 651–4. https://doi.org/10.1007/s10336-008-0291-0

Knox, A.G. (1990) The sympatric breeding of Common and Scottish Crossbills *Loxia curvirostra* and *L. scotica* and the evolution of crossbills. *Ibis* 132: 454–66. https://doi.org/10.1111/j.1474-919X.1990.tb01063.x

Knox, A.G. (1992) Species and pseudospecies: the structure of crossbill populations. *Biological Journal of the Linnean Society* 47: 325–35. https://doi.org/10.1111/j.1095-8312.1992.tb00673.x

Koenig, W.D. (2001) Synchrony and periodicity of eruptions by boreal birds. *Condor* 103: 725–35. https://doi.org/10.1093/condor/103.4.725

Koenig, W.D. and J. Knops, J.M.H. (2000) Patterns of annual seed production by northern hemisphere trees: a global perspective. *American Naturalist* 155: 59–69. https://doi.org/10.1086/303302

Koenig, W.D. and Knops, J.M.H. (2001) Seed-crop size and eruptions of North American boreal seed-eating birds. *Journal of Animal Ecology* 70: 609–20. https://doi.org/10.1046/j.1365-2656.2001.00516.x

Konishi, M., Emlen, S.T., Ricklefs, R.E. and Wingfield, J.C. (1989) Contributions of bird studies to biology. *Science* 246: 465–72. https://doi.org/10.1126/science.2683069

Koprowski, J.L., Ramos, N.S., Pasch, B.S. and Zugmeyer, C.A. (2006) Observations on the ecology of the endemic Mearns's Squirrel (*Tamiasciurus mearnsi*). *Southwestern Naturalist* 51: 426–30. https://doi.org/10.1894/0038-4909(2006)51[426:OOTEOT]2.0.CO;2

Koyabu, D.B., Oshida, T., Dang, N.X., Can, D.N., Kimura, J., Sasaki, M., Motokawa, M., Son, N.T., Hayashida, A., Shintaku, Y. and Endo, H. (2009) Craniodental mechanics and the feeding ecology of two sympatric callosciurine squirrels in Vietnam. *Journal of Zoology* 279: 372–80. https://doi.org/10.1111/j.1469-7998.2009.00629.x

Krause, J. and Ruxton, G.D. (2002) *Living in Groups*. Oxford: Oxford University Press. https://doi.org/10.1093/oso/9780198508175.001.0001

Lack, D. (1944a) Ecological aspects of species-formation in passerine birds. *Ibis* 86: 260–86. https://doi.org/10.1111/j.1474-919X.1944.tb04092.x

Lack, D. (1944b) Correlation between beak and food in the crossbill, *Loxia curvirostra* Linnaeus. *Ibis* 86: 552–3. https://doi.org/10.1111/j.1474-919X.1944.tb02223.x

Lamont, B.B. and Enright, N.J. (2000) Adaptive advantages of aerial seed banks. *Plant Species Biology* 15: 157–66. https://doi.org/10.1046/j.1442-1984.2000.00036.x

Lamont, B.B., Le Maitre, D.C., Cowling, R.M. and Enright, N.J. (1991) Canopy seed storage in woody plants. *The Botanical Review* 57: 277–317. https://doi.org/10.1007/BF02858770

LaMontagne, J.M., Pearse, I.S., Greene, D.F. and Koenig, W.D. (2020) Mast seeding patterns are asynchronous at a continental scale. *Nature Plants* 6: 460–5. https://doi.org/10.1038/s41477-020-0647-x

Landauer, W. (1938) Notes on cross-beak in fowl. *Journal of Genetics* 37: 51–68. https://doi.org/10.1007/BF02982143

Langmore, N.E., Grealy, A., Noh, H.-J., Medina, I., Skeels, A., Grant, J., Murray, K.D., Kilner, R.M., Holleley, C.E. and Holleley, C.E. (2024) Coevolution with hosts underpins speciation in brood-parasitic cuckoos. *Science* 384: 1030–6. https://doi.org/10.1126/science.adj3210

Lauck, K.S., Ke, A., Olimpi, E.M., Paredes, D., Hood, K., Phillips, T., Anderegg, W.R.L. and Karp, D.S. (2023) Agriculture and hot temperatures interactively erode the nest success of habitat generalist birds across the United States. *Science* 382: 290–4. https://doi.org/10.1126/science.add2915

Lavin, S.R., Karasov, W.H., Ives, A.R., Middleton, K.M. and Garland Jr., T. (2008) Morphometrics of the avian small intestine compared with that of nonflying mammals: a phylogenetic approach. *Physiological and Biochemical Zoology* 81: 526–50. https://doi.org/10.1086/590395

Leopold, A. (1987) Foreword. In J.B. Callicott, ed., *Companion to A Sand County Almanac: Interpretative and Critical Essays*, pp. 281–8. Madison: University of Wisconsin Press.

Lesieur, V., Yart, A., Guilbon, S., Lorme, P., Auger-Rozenberg, M.-A. and Roques, A. (2014) The invasive *Leptoglossus* seed bug, a threat for commercial seed crops, but for conifer diversity? *Biological Invasions* 16: 1833–49. https://doi.org/10.1007/s10530-013-0630-9

Lesk, C., Coffel, E., D'Amato, A.W., Dodds, K. and Horton, R. (2017) Threats to North American forests from Southern Pine Beetle with warming winters. *Nature Climate Change* 7: 713–18. https://doi.org/10.1038/nclimate3375

Leslie, A.B. (2011) Predation and protection in the macroevolutionary history of conifer cones. *Proceedings of the Royal Society of London B* 278: 3003–8. https://doi.org/10.1098/rspb.2010.2648

Leslie, A.B., Beaulieu, J.M., Rai, H.S., Crane, P.R., Donoghue, M.J. and Mathews, S. (2012) Hemisphere-scale differences in conifer evolutionary dynamics. *Proceedings of the National Academy of Sciences USA* 109: 16217–21. https://doi.org/10.1073/pnas.1213621109

Lifjeld, J.T, Laskemoen, T., Kleven, O., Albrecht, T. and Robertson, R.J. (2010) Sperm length variation as a predictor of extrapair paternity in passerine birds. *PLoS ONE* 5: e13456–59. https://doi.org/10.1371/journal.pone.0013456

Lima, S.L. (1985) Sampling behavior of starlings in simple patchy environments. *Behavioral Ecology and Sociobiology* 16: 135–42. https://doi.org/10.1007/BF00295147

Linhart, Y.B. (1978) Maintenance of variation in cone morphology in California closed cone pines: the roles of fire, squirrels and seed output. *Southwestern Naturalist* 23: 29–40. https://doi.org/10.2307/3669977

Linhart, Y.B., Moreira, X., Snyder, M.A. and Mooney, K.A. (2014) Variability in seed cone production and functional response of seed predators to seed cone availability: support for the predator satiation hypothesis. *Journal of Ecology* 102: 576–83. https://doi.org/10.1111/1365-2745.12231

Lotan, J.E. (1975) The role of cone serotiny in Lodgepole Pine forests. In D.M. Baumgartner, ed., *Management of Lodgepole Pine Ecosystems*, pp. 471–95. Symposium Proceedings. Washington State University, Pullman, WA.

Low, T. (2014) *Where Song Began: Australia's Birds and How They Changed the World*. New Haven: Yale University Press.

MacDonald, G.M. and Cwynar, L.C. (1985) A fossil pollen based reconstruction of the late Quaternary history of Lodgepole Pine (*Pinus contorta* ssp. *latifolia*) in the western interior of Canada. *Canadian Journal of Forest Research* 15: 1039–44. https://doi.org/10.1139/x85-168

Mackey, B., Campbell, C., Norman, P., Hugh, S., Della Sala, D.A., Malcolm, J.R., Desrochers, M. and Drapeau, P. (2023) Assessing the cumulative impacts of forest management on forest age structure development and woodland caribou habitat in boreal landscapes: a case study from two Canadian provinces. *Land* 13: 6. https://doi.org/10.3390/land13010006

Marlin, K.F., Greene, D.F., Kane, J.M., Reilly, M. and Madurapperuma, B.D. (2024) Evaluating immaturity risk in young stands of the serotinous Knobcone Pine (*Pinus attenuata*). *Ecosphere* 15: e4765. https://doi.org/10.1002/ecs2.4765

Marquiss, M. and Rae, R. (1994) Seasonal trends in abundance, diet and breeding of Common Crossbills (*Loxia curvirostra*) in an area of mixed species conifer plantation following the 1990 crossbill 'irruption'. *Forestry* 67: 31–47. https://doi.org/10.1093/forestry/67.1.31

Marquiss, M. and Rae, R. (2002) Ecological differentiation in relation to bill size amongst sympatric, genetically undifferentiated crossbills *Loxia* spp. *Ibis* 144: 494–508. https://doi.org/10.1046/j.1474-919X.2002.00041.x

Marquiss, M., Newton, I., Hobson, K.A. and Kolbeinsson, Y. (2012) Origins of irruptive migrations by Common Crossbills *Loxia curvirostra* into northwestern Europe revealed by stable isotope analysis. *Ibis* 154: 400–9. https://doi.org/10.1111/j.1474-919X.2012.01221.x

Martin, A.C., Zim, H.S. and Nelson, A.L. (1951) *American Wildlife and Plants*. New York: Dover.

Martin, R., Rochefort, J., Mundry, R. and Segelbacher, G. (2019) Delimitation of call types of Red Crossbill (*Loxia curvirostra*) in the Western Palearctic. *Écoscience* 26: 177–94. https://doi.org/10.1080/11956860.2018.1564483

Martin, R., Rochefort, J., Mundry, R. and Segelbacher, G. (2020) On the relative importance of ecology and geographic isolation as drivers for differentiation of call types of Red Crossbill *Loxia curvirostra* in the Palearctic. *Journal of Avian Biology* 51: 773–15. https://doi.org/10.1111/jav.02358

Martin, R., Rochefort, J., Mundry, R. and Segelbacher, G. (2024) Fast cultural evolution of crossbill (*Loxia* spp.) calls in the Palaearctic. *Ibis* 166: 267–84. https://doi.org/10.1111/ibi.13253

Matziris, D. (1998) Genetic variation in cone and seed characteristics in a clonal seed orchard of Aleppo Pine grown in Greece. *Silvae Genetica* 47: 37–41.

Mayr, E. (1947) Ecological factors in speciation. *Evolution* 1: 263–88. https://doi.org/10.2307/2405327

Mayr, E. (1957) New species of birds described from 1941 to 1955. *Journal für Ornithologie* 98: 22–35. https://doi.org/10.1007/BF01677166

Mayr, E. (1963) *Animal Species and Evolution*. Cambridge: Harvard University Press. https://doi.org/10.4159/harvard.9780674865327

Mayr, E. (1971) New species of birds described from 1956 to 1965. *Journal für Ornithologie* 112: 302–16. https://doi.org/10.1007/BF01640689

Mayr, E. and Vuilleumier, F. (1983) New species of birds described from 1966 to 1975. *Journal für Ornithologie* 124: 217–32. https://doi.org/10.1007/BF01640607

McGinley, M.A., Smith, C.C., Elliott, P.F. and Higgins, J.J. (1990) Morphological constraints on seed mass in Lodgepole Pine. *Functional Ecology* 4: 183–92. https://doi.org/10.2307/2389337

Mehringer Jr., P.J. (1985) Late-Quaternary pollen record from the interior Pacific Northwest and northern Great Basin of the United States. In W.M. Bryant Jr. and R.G. Holloway, eds, *Pollen Records of Late-Quaternary North America Sediments*, pp. 167–89. Dallas: American Association of Stratigraphy and Palynology Foundation.

Mezquida, E.T. and Benkman, C.W. (2005) The geographic selection mosaic for squirrels, crossbills and Aleppo Pine. *Journal of Evolutionary Biology* 18: 348–57. https://doi.org/10.1111/j.1420-9101.2004.00846.x

Mezquida, E.T. and Benkman, C.W. (2010) Habitat area and structure affect the impact of seed predators and the potential for coevolutionary arms races. *Ecology* 91: 802–14. https://doi.org/10.1890/09-0161.1

Mezquida, E.T. and Benkman, C.W. (2014) Causes of variation in biotic interaction strength and phenotypic selection along an altitudinal gradient. *Evolution* 68: 1710–21. https://doi.org/10.1111/evo.12394

Mezquida, E.T. and Benkman, C.W. (2021) Forest and cone structure influence where crossbills forage in a managed Scots Pine forest. *Forest Ecology and Management* 498: 119560. https://doi.org/10.1016/j.foreco.2021.119560

Mezquida, E.T., Svenning, J.-C., Summers, R.W. and Benkman, C.W. (2018) Higher spring temperatures increase food scarcity and limit the current and future distributions of crossbills. *Diversity and Distributions* 24: 473–84. https://doi.org/10.1111/ddi.12694

Mielke, M. and Wassenbergh, S.V. (2022) Three-dimensional movement of the beak during seed processing in domestic canaries. *Journal of Experimental Biology* 225: jeb244360. https://doi.org/10.1242/jeb.244360

Miller, G.E. (1986) Insects and conifer seed production in the Inland Mountain West: a review. In R.C. Shearer, ed., *Proceedings of the Conifer Tree Seed in the Inland Mountain West Symposium*, pp. 225–37. General Technical Report INT 203, United States Department of Agriculture, Forest Service, Ogden, UT.

Mirov, N.T. (1967) *The Genus* Pinus. New York: The Ronald Press Company.

Møller, A.P. (2000) Male parental care, female reproductive success, and extrapair paternity. *Behavioral Ecology* 11: 161–8. https://doi.org/10.1093/beheco/11.2.161

Møller, A.P. and Briskie, J.V. (1995) Extra-pair paternity, sperm competition and the evolution of testis size in birds. *Behavioral Ecology and Sociobiology* 36: 357–65. https://doi.org/10.1007/BF00167797

Mooney, K.A., Linhart, Y.B. and Snyder, M.A. (2010) Masting in ponderosa pine: comparisons of pollen and seed over space and time. *Oecologia* 165: 651–661. https://doi.org/10.1007/s00442-010-1742-x

Morgan, A.D., Gandon, S. and Buckling, A. (2005) The effect of migration on local adaptation in a coevolving host–parasite system. *Nature* 437: 253–6. https://doi.org/10.1038/nature03913

Munar-Delgado, G., Pulido, F. and Edelaar, P. (2024) Performance-based habitat choice can drive rapid adaptive divergence and reproductive isolation. *Current Biology* 34: 5564–69. https://doi.org/10.1016/j.cub.2024.10.006

Naef-Daenzer, B. and Grüebler, M.U. (2016) Post-fledging survival of altricial birds: ecological determinants and adaptation. *Journal of Field Ornithology* 87: 227–50. https://doi.org/10.1111/jofo.12157

Nathan, R., Safriel, U.N., Meir, I.N. and Schiller, G. (1999) Seed release without fire in *Pinus halepensis*, a Mediterranean serotinous wind-dispersed tree. *Journal of Ecology* 87: 659–69. https://doi.org/10.1046/j.1365-2745.1999.00382.x

Naumann, G., Alfieri, L., Wyser, K., Mentaschi, L., Betts, R.A., Carrao, H., Spinoni, J., Vogt, J. and Feyen, L. (2018) Global changes in drought conditions under different levels of warming. *Geophysical Research Letters* 45: 3285–96. https://doi.org/10.1002/2017GL076521

Nethersole-Thompson, D. (1975) *Pine Crossbills: A Scottish Contribution*. Berkhamsted: T. & A.D. Poyser.

Newsome, R.D., and Dix, R.L. (1968) The forests of the Cypress Hills, Alberta and Saskatchewan, Canada. *American Midland Naturalist* 80: 118–85. https://doi.org/10.2307/2423608

Newton, I. (1967) The adaptive radiation and feeding ecology of some British finches. *Ibis* 109: 33–96. https://doi.org/10.1111/j.1474-919X.1967.tb00005.x

Newton, I. (1972) *Finches*. London: Collins.

Newton, I. (2006) Movement patterns of Common Crossbills *Loxia curvirostra* in Europe. *Ibis* 148: 782–8. https://doi.org/10.1111/j.1474-919X.2006.00585.x

Newton, I. and Marquiss, M. (1982) Food, predation and breeding season in Sparrowhawks (*Accipiter nisus*). *Journal of Zoology* 197: 221–40. https://doi.org/10.1111/jzo.1982.197.2.221

Nosil, P. (2012) *Ecological Speciation*. Oxford: Oxford University Press. https://doi.org/10.1093/acprof:osobl/9780199587100.001.0001

Nosil, P., Vines, T.H. and Funk, D.J. (2005) Perspective: reproductive isolation caused by natural selection against immigrants from divergent habitats. *Evolution* 59: 705–19. https://doi.org/10.1111/j.0014-3820.2005.tb01747.x

Nuijens, F.W. and Bout, R.G. (1998) The role of two jaw ligaments in the evolution of passerines. *Zoology* 101: 24–33.

Nuijens, F.W. and Zweers, G.A. (1997) Characters discriminating two seed husking mechanisms in finches (Fringillidae: Carduelinae) and estrildids (Passeridae: Estrildinae). *Journal of Morphology* 232: 1–33. https://doi.org/10.1002/(SICI)1097-4687(199704)232:1<1::AID-JMOR1>3.0.CO;2-G

Ockendon, N., Baker, D.J., Carr, J.A., White, E.C., Almond, R.E.A., Amano, T., Bertram, E., Bradbury, R.B., Bradley, C., Butchart, S.H.M., Doswald, N., Foden, W., Gill, D.J.C., Green, R.E., Sutherland, W.J., Tanner, E.V.J. and Pearce-Higgins, J.W. (2014) Mechanisms underpinning climatic impacts on natural populations: altered species interactions are more important than direct effects. *Global Change Biology* 20: 2221–9. https://doi.org/10.1111/gcb.12559

Päckert, M, Favre, A., Schnitzler, J., Martens, J., Sun, Y-H., Tietze, D.T., Hailer, F., Michalak, I. and Strutzenberger, P. (2020) 'Into and out of' the Qinghai-Tibet Plateau and the Himalayas: centers of origin and diversification across five clades of Eurasian montane and alpine passerine birds. *Ecology and Evolution* 10: 9283–300. https://doi.org/10.1002/ece3.6615

Palmer, A.R. (2004) Symmetry breaking and the evolution of development. *Science* 306: 828–33. https://doi.org/10.1126/science.1103707

Palmer, A.R. (2016) What determines direction of asymmetry: genes, environment or chance? *Philosophical Transactions of the Royal Society of London B* 371: 20150417-16. https://doi.org/10.1098/rstb.2015.0417

Parchman, T.L. and Benkman, C.W. (2002) Diversifying coevolution between crossbills and Black Spruce on Newfoundland. *Evolution* 56: 1663–72. https://doi.org/10.1111/j.0014-3820.2002.tb01478.x

Parchman, T.L. and Benkman, C.W. (2008) The geographic selection mosaic for Ponderosa Pine and crossbills: a tale of two squirrels. *Evolution* 62: 348–60. https://doi.org/10.1111/j.1558-5646.2007.00295.x

Parchman, T.L., Benkman, C.W. and Britch, S.C. (2006) Patterns of genetic variation in the adaptive radiation of New World crossbills (Aves: *Loxia*). *Molecular Ecology* 15: 1873–87. https://doi.org/10.1111/j.1365-294x.2006.02895.x

Parchman, T.L., Benkman, C.W. and Mezquida, E.T. (2007) Coevolution between Hispaniolan Crossbills and pine: Does more time allow for greater phenotypic escalation at lower latitude? *Evolution* 61: 2142–53. https://doi.org/10.1111/j.1558-5646.2007.00172.x

Parchman, T.L., Buerkle, C.A., Soria-Carrasco, V. and Benkman, C.W. (2016) Genome divergence and diversification within a geographic mosaic of coevolution. *Molecular Ecology* 25: 5705–18. https://doi.org/10.1111/mec.13825

Parchman, T.L., Edelaar, P., Uckele, K., Mezquida, E.T., Alonso, D., Jahner, J.P., Summers, R.W. and Benkman, C.W. (2018) Resource stability and geographic isolation are associated with genome divergence in western Palearctic crossbills. *Journal of Evolutionary Biology* 31: 1715–31. https://doi.org/10.1111/jeb.13367

Parks, S.A. and Abatzoglou, J.T. (2020) Warmer and drier fire seasons contribute to increases in area burned at high severity in western US forests from 1985 to 2017. *Geophysical Research Letters* 47: e2020GL089858. https://doi.org/10.1029/2020GL089858

Pauw, A., Cocucci, A.A. and Sérsic, A.N. (2020) The least effective pollinator principle: specialized morphology despite generalized ecology. *Plant Biology* 22: 924–31. https://doi.org/10.1111/plb.13145

Payne, R.B. (1972) Nuts, bones, and a nesting of Red Crossbills in the Panamint Mountains, California. *Condor* 74: 485–6. https://doi.org/10.2307/1365913

Payne, R.B. (1987) Populations and type specimens of a nomadic bird: comments on the North American crossbills *Loxia pusilla* Gloger 1834 and *Crucirostra minor* Brehm 1845. *Occasional Papers of the Museum of Zoology* 714: 1–40.

Pearce-Higgins, J.W. and Green, R.E. (2014) *Birds and Climate Change: Impacts and Conservation Responses.* Cambridge: Cambridge University Press. https://doi.org/10.1017/CBO9781139047791

Penny, D. (2001) A 40,000 year palynological record from north-east Thailand; implications for biogeography and palaeo-environmental reconstruction. *Palaeogeography, Palaeoclimatology, Palaeoecology* 171: 97–128. https://doi.org/10.1016/S0031-0182(01)00242-5

Pepper, J.W., Male, T.D. and Roberts, G.E. (2000) Foraging ecology of the South Australian Glossy Black-Cockatoo (*Calyptorhynchus lathami halmaturinus*). *Austral Ecology* 25: 16–24. https://doi.org/10.1046/j.1442-9993.2000.01018.x

Perry, D.A. and Lotan, J.E. (1977) Opening temperatures in serotinous cones of Lodgepole Pine. USDA Forest Service Research Note INT-228.

Peterson, R.T. and Fisher, J. (1955) *Wild America.* New York: Houghton Mifflin.

Petrie, M.D., Hubbard, R.M., Bradford, J.B., Kolb, T.E., Noel, A., Schlaepfer, D.R., Bowen, M.A., Fuller, L.R. and Moser, W.K. (2023) Widespread regeneration failure in Ponderosa Pine forests of the southwestern United States. *Forest Ecology and Management* 545: 121208. https://doi.org/10.1016/j.foreco.2023.121208

Petty, S.J., Patterson, I.J., Anderson, D., Little, B. and Davison, M. (1995) Numbers, breeding performance, and diet of the Sparrowhawk *Accipiter nisus* and Merlin *Falco columbarius* in relation to cone crops and seed-eating finches. *Forest Ecology and Management* 79: 133–46. https://doi.org/10.1016/0378-1127(95)03624-5

Piertney, S.B., Summers, R.W. and Marquiss, M. (2001) Microsatellite and mitochondrial DNA homogeneity among phenotypically diverse crossbill taxa in the UK. *Proceedings of the Royal Society of London B* 268: 1511–17. https://doi.org/10.1098/rspb.2001.1015

Poisot, T., Bever, J.D., Nemri, A., Thrall, P.H. and Hochberg, M.E. (2011) A conceptual framework for the evolution of ecological specialisation. *Ecology Letters* 14: 841–51. https://doi.org/10.1111/j.1461-0248.2011.01645.x

Pomeroy, D.E. (1962). Birds with abnormal bills. *British Birds* 55: 49–72.

Porter, C.K. and Benkman, C.W. (2019) Character displacement of a learned behaviour and its implications for ecological speciation. *Proceedings of the Royal Society of London B* 286: 20190761-10. https://doi.org/10.1098/rspb.2019.0761

Porter, C.K. and Benkman, C.W. (2022) Performance tradeoffs and resource availability drive variation in reproductive isolation between sympatrically diverging crossbills. *American Naturalist* 199: 362–79. https://doi.org/10.1086/718235

Porter, C.K. and Smith, J.W. (2020) Diversification in trophic morphology and a mating signal are coupled in the early stages of sympatric divergence in crossbills. *Biological Journal of the Linnean Society* 129: 74–87. https://doi.org/10.1093/biolinnean/blz163

Porter, C.K., Golcher-Benavides, J. and Benkman, C.W. (2022) Seasonal patterns of dietary partitioning in vertebrates. *Ecology Letters* 25: 2463–75. https://doi.org/10.1111/ele.14100

Prévost, Y.H., Laing, J.E. and Haavisto, V.F. (1988) Seasonal damage by insects and squirrels to female reproductive structures of Black Spruce, *Picea mariana* (Mill.) B.S.P. *Canadian Entomologist* 1(20): 1113–21. https://doi.org/10.4039/Ent1(201113-12

Price, T. (2008) *Speciation in Birds.* Greenwood Village: Roberts & Company Publishers.

Pringle, R.M. (2023) Wolves, otters, ungulates, and a promising path for ecology. *Proceedings of the National Academy of Sciences USA* 120: e2221817120. https://doi.org/10.1073/pnas.2221817120

Pringle, R.M. and Hutchinson, M.C. (2020) Resolving food-web structure. *Annual Review of Ecology, Evolution and Systematics* 51: 55–80. https://doi.org/10.1146/annurev-ecolsys-110218-024908

Pulliam, H.R. (1975) Coexistence of sparrows: a test of community theory. *Science* 189: 474–6. https://doi.org/10.1126/science.189.4201.474

Questiau, S., Gielly, L., Clouet, M. and Taberlet, P. (1999) Phylogeographical evidence of gene flow among Common Crossbill (*Loxia curvirostra*, Aves, Fringillidae) populations at the continental level. *Heredity* 83: 196–205. https://doi.org/10.1046/j.1365-2540.1999.00551.x

Raffa, K.F., Aukema, B.H., Bentz, B.J., Carroll, A.L., Hicke, J.A., Turner, M.G. and Romme, W.H. (2008) Cross-scale drivers of natural disturbances prone to anthropogenic amplification: the dynamics of bark beetle eruptions. *BioScience* 58: 501–17. https://doi.org/10.1641/B580607

Ramsey, J., Bradshaw Jr., H.D. and Schemske, D.W. (2003) Components of reproductive isolation between the monkeyflowers *Mimulus lewisii* and *M. cardinalis* (Phyrmaceae). *Evolution* 57: 1520–34. https://doi.org/10.1111/j.0014-3820.2003.tb00360.x

Renner, H.M., Piatt, J.F., Renner, M., Drummond, B.A., Laufenberg, J.S. and Parrish, J.K. (2024) Catastrophic and persistent loss of Common Murres after a marine heatwave. *Science* 386: 1272–6. https://doi.org/10.1126/science.adq4330

Reudink, M.W., Mech, S.G. and Curry, R.L. (2006) Extrapair paternity and mate choice in a chickadee hybrid zone. *Behavioral Ecology* 17: 56–62. https://doi.org/10.1093/beheco/ari098

Rexstad, E. (1982) *Bird Model, Version III: Description and Documentation.* Logan: Utah State University.

Robb, M.S. (2000) Introduction to vocalizations of crossbills in north-western Europe. *Dutch Birds* 22: 61–107.

Robinson, B.W. and Wilson, D.S. (1998) Optimal foraging, specialization, and a solution to Liem's Paradox. *American Naturalist* 151: 223–35. https://doi.org/10.1086/286113

Robledo-Arnuncio, J.J., Alı'a, R. and Gil, L. (2004) Increased selfing and correlated paternity in a small population of a predominantly outcrossing conifer, *Pinus sylvestris*. *Molecular Ecology* 13: 2567–77. https://doi.org/10.1111/j.1365-294X.2004.02251.x

Rundle, H.D., Nagel, L., Boughman, J.W. and Schluter, D. (2000) Natural selection and parallel speciation in sympatric sticklebacks. *Science* 287: 306–8. https://doi.org/10.1126/science.287.5451.306

Santisteban, L., Benkman, C.W., Fetz, T. and Smith, J.W. (2012) Survival and population size of a resident bird species are declining as temperature increases. *Journal of Animal Ecology* 81: 352–63. https://doi.org/10.1111/j.1365-2656.2011.01918.x

Sauchyn, D.J. (1997) Proxy records of postglacial climate in the Canadian Prairie Provinces: a guide to the literature and current research. Appendix I. In R. Herrington, B. Johnson and F. Hunter, eds, *Responding to Climate Change in the Prairies: Volume Three of the Canada Country Study: Climate Change and Adaptation*. Ottawa: Environment Canada.

Sauchyn, M.A. and Sauchyn, D.J. (1991) A continuous record of Holocene pollen from Harris Lake, southwestern Saskatchewan, Canada. *Palaeogeography, Palaeoclimatology, Palaeoecology* 88: 13–23. https://doi.org/10.1016/0031-0182(91)90012-G

Sauer, J.R., Hines, J.E., Fallon, J.E., Pardieck, K.L., Ziolkowski Jr., D.J. and Link, W.A. (2014) *The North American Breeding Bird Survey, Results and Analysis 1966-2013*. Version 01.30.2015, USGS Patuxent Wildlife Research Center, Laurel, MD. http://www.mbr-pwrc.usgs. gov/bbs/

Schaller, G.B. (2012) *Tibet Wild: A Naturalist's Journeys on the Roof of the World.* Washington, DC: Island Press. https://doi.org/10.5822/978-1-61091-232-7_9

Schluter, D. (1996) Ecological causes of adaptive radiation. *American Naturalist* 148: S40–64. https://doi.org/10.1086/285901

Schluter, D. (2001) Ecology and the origin of species. *Trends in Ecology and Evolution* 16: 372–80. https://doi.org/10.1016/S0169-5347(01)02198-X

Schluter, D. (2009) Evidence for ecological speciation and its alternative. *Science* 323: 737–41. https://doi.org/10.1126/science.1160006

Schluter, D. (2010) Resource competition and coevolution in sticklebacks. *Evolution: Education and Outreach* 3: 54–61. https://doi.org/10.1007/s12052-009-0204-6

Schluter, D. and Grant, P.R. (1984) Determinants of morphological patterns in communities of Darwin's finches. *American Naturalist* 123: 175–96. https://doi.org/10.1086/284196

Schluter, D. and McPhail, J.D. (1992) Ecological character displacement and speciation in sticklebacks. *American Naturalist* 140: 85–108. https://doi.org/10.1086/285404

Schoennagel, T., Turner, M.G. and Romme, W.H. (2003). The influence of fire interval and serotiny on postfire Lodgepole Pine density in Yellowstone National Park. *Ecology* 84: 2967–78. https://doi.org/10.1890/02-0277

Schoettle, A.W. and Sniezko, R.A. (2007) Proactive intervention to sustain high-elevation pine ecosystems threatened by White Pine Blister Rust. *Journal of Forest Research* 12: 327–36. https://doi.org/10.1007/s10310-007-0024-x

Schondube, J.E. and Martinez del Rio, C. (2003) The flowerpiercers' hook: an experimental test of an evolutionary trade-off. *Proceedings of the Royal Society of London B* 270: 195–8. https://doi.org/10.1098/rspb.2002.2231

Schopmeyer, C.S. (1974) *Seeds of Woody Plants of the United States*. Agriculture Handbook 450. USDA Forest Service, Washington, DC.

Schroeder, M.A., Young, J.R. and Braun, C.E. (2020) Greater Sage-Grouse (*Centrocercus urophasianus*), version 1.0. In A.F. Poole and F.B. Gill, eds, *Birds of the World*. Cornell Lab of Ornithology, Ithaca, NY. https://doi.org/10.2173/bow.saggro.01

Schulte, L., Li, C., Lisovski, S. and Herzschuh, U. (2022) Forest-permafrost feedbacks and glacial refugia help explain the unequal distribution of larch across continents. *Journal of Biogeography* 49: 1825–38. https://doi.org/10.1111/jbi.14456

Sekercioglu, C.H., Wenny, D.G. and Whelan, C.J. (2016) *Why Birds Matter: Avian Ecological Function and Ecosystem Services*. Chicago: University of Chicago Press. https://doi.org/10.7208/chicago/9780226382777.001.0001

Senar, J.C., Borras, A., Cabrera, J. and Cabrera, T. (1995) The bill crossover ratio of the Common Crossbill, *Loxia curvirostra*, is not always 1:1. *Ardeola* 42: 209–10.

Senar, J.C., Mateos-Gonzalez, F., Uribe, F. and Arroyo, L. (2013) Familiarity adds to attractiveness in matters of siskin mate choice. *Proceedings of the Royal Society of London B* 280: 20132361. https://doi.org/10.1098/rspb.2013.2361

Sewall, K.B. (2009) Limited adult learning maintains call dialects but permits pair-distinctive calls in Red Crossbills. *Animal Behaviour* 77: 1303–11. https://doi.org/10.1016/j.anbehav.2009.01.033

Sewall, K.B. (2011) Early learning of discrete call variants in Red Crossbills: implications for reliable signaling. *Behavioral Ecology and Sociobiology* 65: 157–66. https://doi.org/10.1007/s00265-010-1022-0

Shaw, J.J., Tregenza, T., Parker, G.A. and Harvey, I.A. (1995) Evolutionarily stable foraging speeds in feeding scrambles: a model and an experimental test. *Proceedings of the Royal Society of London B* 260: 273–7. https://doi.org/10.1098/rspb.1995.0091

Sibley, D.A. (2014) *The Sibley Guide to Birds*, 2nd edn. New York: Alfred A. Knopf.

Siepielski, A.M. and Benkman, C.W. (2005) A role for habitat area in the geographic mosaic of coevolution between Red Crossbills and Lodgepole Pine. *Journal of Evolutionary Biology* 18: 1042–9. https://doi.org/10.1111/j.1420-9101.2005.00902.x

Siepielski, A.M. and Benkman, C.W. (2007) Convergent patterns in the selection mosaic for two North American bird-dispersed pines. *Ecological Monographs* 77: 203–20. https://doi.org/10.1890/06-0929

Silva, M., Brown, J.H. and Downing, J.A. (1997) Differences in population density and energy use between birds and mammals: a macroecological perspective. *Journal of Animal Ecology* 66: 327–40. https://doi.org/10.2307/5979

Singh, N.B. and Chaudhary, V.K. (1993) Variability, heritability and genetic gain in cone and nut characters of Chilgoza Pine (*Pinus gerardiana* Wall.). *Silvae Genetica* 42: 61–3.

Smith, C.C. (1968) The adaptive nature of social organization in the genus of tree squirrels *Tamiasciurus*. *Ecological Monographs* 38: 31–64. https://doi.org/10.2307/1948536

Smith, C.C. (1970) The coevolution of pine squirrels (*Tamiasciurus*) and conifers. *Ecological Monographs* 40: 349–71. https://doi.org/10.2307/1942287

Smith, C.C. (1981) The indivisible niche of *Tamiasciurus*: an example of nonpartitioning of resources. *Ecological Monographs* 51: 343–63. https://doi.org/10.2307/2937278

Smith, C.C. and Balda, R.P. (1979) Competition among insects, birds and mammals for conifer seeds. *American Zoologist* 19: 1065–83. https://doi.org/10.1093/icb/19.4.1065

Smith, C.C., Hamrick, J.L. and Kramer, C.L. (1988) The effects of stand density on frequency of filled seeds and fecundity of Lodgepole Pine (*Pinus contorta* Dougl.). *Canadian Journal of Forest Research* 18: 453–60. https://doi.org/10.1139/x88-066

Smith, C.C., Hamrick, J.L. and Kramer, C.L. (1990) The advantage of mast years for wind pollination. *American Naturalist* 136: 154–66. https://doi.org/10.1086/285089

Smith, J.W. and Benkman, C.W. (2007) A coevolutionary arms race causes ecological speciation in crossbills. *American Naturalist* 169: 455–65. https://doi.org/10.1086/511961

Smith, J.W., Benkman, C.W. and Coffey, K. (1999) The use and misuse of public information by foraging Red Crossbills. *Behavioral Ecology* 10: 54–62. https://doi.org/10.1093/beheco/10.1.54

Smith, J.W., Sjoberg, S.M., Mueller, M.C. and Benkman, C.W. (2012) Assortative flocking in crossbills and implications for ecological speciation. *Proceedings of the Royal Society of London B* 279: 4223–9. https://doi.org/10.1098/rspb.2012.1500

Snowberg, L.K. and Benkman, C.W. (2007) The role of marker traits in the assortative mating within Red Crossbills, *Loxia curvirostra* complex. *Journal of Evolutionary Biology* 20: 1924–32. https://doi.org/10.1111/j.1420-9101.2007.01372.x

Snowberg, L.K. and Benkman, C.W. (2009) Mate choice based on a key ecological performance trait. *Journal of Evolutionary Biology* 22: 762–9. https://doi.org/10.1111/j.1420-9101.2009.01699.x

Sobral, M., Silvius, K.M., Overman, H., Oliveira, L.F.B., Raab, T.K., and Fragoso, J.M.V. (2017) Mammal diversity influences the carbon cycle through trophic interactions in the Amazon. *Nature Ecology and Evolution* 1: 1670–6. https://doi.org/10.1038/s41559-017-0334-0

Sperber, A.L., Werner, L.M., Kappeler, P.M. and Fichtel, C. (2017) Grunt to go—vocal coordination of group movements in Redfronted Lemurs. *Ethology* 123: 894–905. https://doi.org/10.1111/eth.12663

Stamps, J.A. (1988) Conspecific attraction and aggregation in territorial species. *American Naturalist* 131: 329–47. https://doi.org/10.1086/284793

Stevens-Rumann, C., Prichard, S., Whitman, E., Parisien, M. and Meddens, A. (2022) Considering regeneration failure in the context of changing climate and disturbance regimes in western North America. *Canadian Journal of Forest Research* 52: 1281–302. https://doi.org/10.1139/cjfr-2022-0054

Stevenson, J., Siringan, F., Finn, J., Madulid, D. and Heijnis, H. (2010) Paoay Lake, northern Luzon, the Philippines: a record of Holocene environmental change. *Global Change Biology* 16: 1672–88. https://doi.org/10.1111/j.1365-2486.2009.02039.x

Strauss, L. (2001) Fire frequency of the Cypress Hills West Block forest. Unpublished MSc thesis, University of Regina.

Strong, C., Zuckerberg, B., Betancourt, J.L. and Koenig, W.D. (2015) Climatic dipoles drive two principal modes of North American boreal bird irruption. *Proceedings of the National Academy of Sciences USA* 112: E2795802. https://doi.org/10.1073/pnas.1418414112

Sudworth, G.B. (1917) *The Pine Trees of the Rocky Mountain Region.* Washington, DC: United States Department of Agriculture Bulletin no. 460.

Sudworth, G.B. (1967) *Forest Trees of the Pacific Slope*. New York: Dover.

Sullivan, K.A. (1989) Predation and starvation: age-specific mortality in juvenile juncos (*Junco phaenotus*). *Journal of Animal Ecology* 58: 275–86. https://doi.org/10.2307/5000

Summers, R.W. (2002) Parrot Crossbills breeding in Abernethy Forest, Highland. *British Birds* 95: 4–11.

Summers, R.W. (2011) Patterns of exploitation of annually varying *Pinus sylvestris* seed crops by seed-eaters of differing dispersal ability. *Ecography* 34: 723–8. https://doi.org/10.1111/j.1600-0587.2010.06498.x

Summers, R.W. (2018) Foraging patterns of Common Crossbills (*Loxia curvirostra*) on spruces (*Picea* spp.) in Scotland. *Forestry: An International Journal of Forest Research* 91: 444–50. https://doi.org/10.1093/forestry/cpy002

Summers, R.W. (2020) Calls of crossbills in Scotland: association with bird size and changes over time. *British Birds* 113: 152–64.

Summers, R.W. and Proctor, R. (1999) Tree and cone selection by crossbills *Loxia* sp. and Red Squirrels *Sciurus vulgaris* at Abernethy Forest, Strathspey. *Forest Ecology and Management* 118: 173–82. https://doi.org/10.1016/S0378-1127(98)00496-4

Summers, R.W. and Proctor, R. (2005) Timing of shedding seeds and cones, and production in different stands of Scots Pines at Abernethy Forest, Scotland. *Forestry* 78: 541–9. https://doi.org/10.1093/forestry/cpi047

Summers, R.W., Jardine, D.C., Marquiss, M. and Proctor, R. (1996) The biometrics of invading Common Crossbills *Loxia curvirostra* in Britain during 1990–1991. *Ringing and Migration* 17: 1–10. https://doi.org/10.1080/03078698.1996.9674113

Summers, R.W., Jardine, D., Marquiss, M. and Rae, R. (2002) The distribution and habitats of crossbills *Loxia* spp. in Britain, with special reference to the Scottish Crossbill *Loxia scotica*. *Ibis* 144: 393–410. https://doi.org/10.1046/j.1474-919X.2002.00064.x

Summers, R.W., Dawson, R.J.G. and Phillips, R.E. (2007) Assortative mating and patterns of inheritance indicate that the three crossbill taxa in Scotland are species. *Journal of Avian Biology* 38: 153–62. https://doi.org/10.1111/j.2007.0908-8857.03798.x

Summers, R.W., Dawson, R.J.G. and Proctor, R. (2010) Temporal variation in breeding and cone size selection by three species of crossbills *Loxia* spp. in a native Scots pinewood. *Journal of Avian Biology* 41: 219–28. https://doi.org/10.1111/j.1600-048X.2009.04768.x

Summers, R.W., Swallow, B., Fridman, J., Hokkanen, T., Newton, I. and Buckland, S.T. (2024) Irruptions of crossbills *Loxia* spp. in northern Europe—patterns and correlations with seed production by key and non-key conifers. *Ibis* 166: 1172–83. https://doi.org/10.1111/ibi.13328

Sumpter, D.J.T. and Pratt, S.C. (2009) Quorum responses and consensus decision making. *Philosophical Transactions of the Royal Society London B* 364: 743–53. https://doi.org/10.1098/rstb.2008.0204

Svärdson, G. (1957) The 'invasion' type of bird migration. *British Birds* 50: 314–43.

Symonds, M.R.E. and Tattersall, G.J. (2010) Geographical variation in bill size across bird species provides evidence for Allen's Rule. *American Naturalist* 176: 188–97. https://doi.org/10.1086/653666

Taber, W. (1968) Life histories of North American cardinals, grosbeaks, buntings, towhees, finches, sparrows, and allies, Part 1, ed. O.L. Austin Jr. *United States National Museum Bulletin* 237. https://doi.org/10.5479/si.03629236.237.1

Talluto, M.V. and Benkman, C.W. (2013) Landscape-scale eco-evolutionary dynamics: selection by seed predators and fire determine a major reproductive strategy. *Ecology* 94: 1307–16. https://doi.org/10.1890/12-2058.1

Talluto, M.V. and Benkman, C.W. (2014) Conflicting selection from fire and seed predation drives fine-scaled phenotypic variation in a widespread North American conifer. *Proceedings of the National Academy of Sciences USA* 111: 9543–8. https://doi.org/10.1073/pnas.1400944111

Tattersall, G.J., Arnaout, B. and Symonds, M.R.E. (2017) The evolution of the avian bill as a thermoregulatory organ. *Biological Reviews* 92: 1630–56. https://doi.org/10.1111/brv.12299

Temeles, E.J. and Kress, W.J. (2003) Adaptation in a plant–hummingbird association. *Science* 300: 630–3. https://doi.org/10.1126/science.1080003

ten Cate, C. and Rowe, C. (2007) Biases in signal evolution: learning makes a difference. *Trends in Ecology and Evolution* 22: 380–7. https://doi.org/10.1016/j.tree.2007.03.006

Thompson, J.N. (1982) *Interaction and Coevolution*. New York: Wiley. https://doi.org/10.7208/chicago/9780226127323.001.0001

Thompson, J.N. (1994) *The Coevolutionary Process*. Chicago: University of Chicago Press.

Thompson, J.N. (1999) Specific hypotheses on the geographic mosaic of coevolution. *American Naturalist* 153: S1–14. https://doi.org/10.1086/303208

Thompson, J.N. (2005) *The Geographic Mosaic of Coevolution*. Chicago: University of Chicago Press. https://doi.org/10.7208/chicago/9780226118697.001.0001

Thompson, J.N. (2013) *Relentless Evolution*. Chicago: University of Chicago Press.

Thompson, J.N. and Fernandez, C.C. (2006) Temporal dynamics of antagonism and mutualism in a geographically variable plant–insect interaction. *Ecology* 87: 103–12. https://doi.org/10.1890/05-0123

Toenies, M.J., Miller, D.A.W., Marshall, M.R. and Stauffer, G.E. (2018) Shifts in vegetation and avian community structure following the decline of a foundational forest species, the Eastern Hemlock. *Ornithological Applications* 120: 489–506. https://doi.org/10.1650/CONDOR-17-204.1

Tomback, D.F. (1982) Dispersal of Whitebark Pine seeds by Clark's Nutcracker: a mutualism hypothesis. *Journal of Animal Ecology* 51: 451–67. https://doi.org/10.2307/3976

Tombre-Steen, I. (1991) Foraging behaviour in the Parrot Crossbill *Loxia pytyopsittacus*: systematic searching in patchy environments. *Ornis Scandinavica* 22: 383–6. https://doi.org/10.2307/3676512

Tordoff, H.B. (1954) Social organization and behavior in a flock of captive, nonbreeding Red Crossbills. *Condor* 56: 346–58. https://doi.org/10.2307/1365020

Tordoff, H.B., and Dawson, W.R. (1965) The influence of daylength on reproductive timing in the Red Crossbill. *Condor* 67: 416–422. https://doi.org/10.2307/1365634

Turbek, S.P., Browne, M., Giacomo, A.S.D., Kopuchian, C., Hochachka, W.M., Estalles, C., Lijtmaer, D.A., Tubaro, P.L., Silveira, L.F., Lovette, I.J., Safran, R.J., Taylor, S.A. and Campagna, L. (2021) Rapid speciation via the evolution of pre-mating isolation in the Iberá Seedeater. *Science* 371: eabc0256. https://doi.org/10.1126/science.abc0256

Turgeon, J.J., Roques, A. and de Groot, P. (1994) Insect fauna of coniferous seed cones: diversity, host plant interactions, and management. *Annual Review of Ecology and Systematics* 39: 179–212. https://doi.org/10.1146/annurev.en.39.010194.001143

Turnbull, J.W., Armitage, F.B. and Burley, J. (1980) Distribution and ecology of the *Pinus kesiya* complex. *Tropical Forestry Papers* 9: 13–45.

Turner, M.G., Romme, W.H. and Tinker, D.B. (2003) Surprises and lessons from the 1988 Yellowstone fires. *Frontiers in Ecology and the Environment* 1: 351–8. https://doi.org/10.1890/1540-9295(2003)001[0351:SALFTY]2.0.CO;2

Turner, M.G., Braziunas, K.H., Hansen, W.D. and Harvey, B.J. (2019) Short-interval severe fire erodes the resilience of subalpine Lodgepole Pine forests. *Proceedings of the National Academy of Sciences USA* 116: 11319–28. https://doi.org/10.1073/pnas.1902841116

Turner, M.G., Braziunas, K.H., Hansen, W.D., Hoecker, T.J., Rammer, W., Ratajczak, Z., Westerling, A.L. and Seidl, R. (2022) The magnitude, direction, and tempo of forest change in Greater Yellowstone in a warmer world with more fire. *Ecological Monographs* 92: e01485. https://doi.org/10.1002/ecm.1485

Urban, M.C. (2024) Climate change extinctions. *Science* 386: 1123–8. https://doi.org/10.1126/science.adp4461

Vallender, R., Friesen, V.L. and Robertson, R.J. (2007) Paternity and performance of Golden-winged Warblers (*Vermivora chrysoptera*) and Golden-winged X Blue-winged Warbler (*V. pinus*) hybrids at the leading edge of a hybrid zone. *Behavioral Ecology and Sociobiology* 61: 1797–807. https://doi.org/10.1007/s00265-007-0413-3

Valone, T.J. (1989) Group foraging, public information, and patch estimation. *Oikos* 56: 357–63. https://doi.org/10.2307/3565621

Valone, T.J. (1993) Patch information and estimation: a cost of group foraging. *Oikos* 68: 258–66. https://doi.org/10.2307/3544838

van der Meij, M.A.A. and Bout, R.G. (2006) Seed husking time and maximal bite force in finches. *Journal of Experimental Biology* 209: 3329–35. https://doi.org/10.1242/jeb.02379

van der Meij, M.A.A. and Bout, R.G. (2008) The relationship between shape of the skull and bite force in finches. *Journal of Experimental Biology* 211: 1668–80. https://doi.org/10.1242/jeb.015289

van Doorn, G.S., Edelaar, P. and Weissing, F.J. (2009) On the origin of species by natural and sexual selection. *Science* 326: 1704–7. https://doi.org/10.1126/science.1181661

van Mantgem, P.J., Stephenson, N.L., Byrne, J.C., Daniels, L.D., Franklin, J.F., Fule, P.Z., Harmon, M.E., Larson, A.J., Smith, J.M., Taylor, A.H. and Veblen, T.T. (2009) Widespread increase of tree mortality rates in the western United States. *Science* 323: 521–4. https://doi.org/10.1126/science.1165000

Vander Wall, S.B. and Balda, R.P. (1977) Coadaptations of the Clark's Nutcracker and the Piñon Pine for efficient seed harvest and dispersal. *Ecological Monographs* 47: 89–111. https://doi.org/10.2307/1942225

Veen, T., Borge, T., Griffith, S.C., Saetre, G.P., Bures, S., Gustafsson, L. and Sheldon, B.C. (2001) Hybridization and adaptive mate choice in flycatchers. *Nature* 411: 45–50. https://doi.org/10.1038/35075000

Veen, T., and Otto, S.P. (2015) Liking the good guys: amplifying local adaptation via the evolution of condition-dependent mate choice. *Journal of Evolutionary Biology* 28: 1804–15. https://doi.org/10.1111/jeb.12696

Viglas, J.N., Brown, C.D. and Johnstone, J.F. (2013) Age and size effects on seed productivity of northern Black Spruce. *Canadian Journal of Forest Research* 43: 534–43. https://doi.org/10.1139/cjfr-2013-0022

Villa, S.M., Altuna, J.C., Ruff, J.S., Beach, A.B., Mulvey, L.I., Poole, E.J., Campbell, H.E., Johnson, K.P., Shapiro, M.D., Bush, S.E. and Clayton, D.H. (2019) Rapid experimental evolution of reproductive isolation from a single natural population. *Proceedings of the National Academy of Sciences USA* 6: 201901247-6. https://doi.org/10.1073/pnas.1901247116

Vuilleumier, F. and Mayr, E. (1987) New species of birds described from 1976 to 1980. *Journal für Ornithologie* 128: 137–50. https://doi.org/10.1007/BF01661691

Walsberg, G.E. (1983) Avian ecological energetics. In D.S. Farner, J.R. King and K.C. Parkes, eds, *Avian Biology*, Vol. 7, pp. 161–20. New York: Academic Press. https://doi.org/10.1016/B978-0-12-249407-9.50012-0

Werner, E.E. (1977) Species packing and niche complementarity in three sunfishes. *American Naturalist* 111: 553–78. https://doi.org/10.1086/283184

Werner, W.L. (1997) Pines and other conifers in Thailand—a Quaternary relic? *Journal of Quaternary Science* 12: 451–4. https://doi.org/10.1002/(SICI)1099-1417(199709/10)12:5<451::AID-JQS328>3.0.CO;2-F

West, R.J. (1989) Cone depredations by the Red Squirrel in Black Spruce stands in Newfoundland: implications for commercial cone collection. *Canadian Journal of Forest Research* 19: 1207–10. https://doi.org/10.1139/x89-182

Westerling, A.L., Hidalgo, H.G., Cayan, D.R. and Swetnam, T.W. (2006) Warming and earlier spring increase western US forest wildfire activity. *Science* 313: 940–3. https://doi.org/10.1126/science.1128834

Westerling, A.L., Turner, M.G., Smithwick, E.A.H., Romme, W.H. and Ryan, M.G. (2011) Continued warming could transform Greater Yellowstone fire regimes by mid-21st century. *Proceedings of the National Academy of Sciences USA* 108: 13165–70. https://doi.org/10.1073/pnas.1110199108

Westoby, M., Leishman, M. and Lord, J. (1996) Comparative ecology of seed size and dispersal. *Philosophical Transactions of the Royal Society of London B* 351: 1309–17. https://doi.org/10.1098/rstb.1996.0114

Wheeler, N.C. and Guries, R.P. (1982) Biogeography of Lodgepole Pine. *Canadian Journal of Botany* 60: 1805–14. https://doi.org/10.1139/b82-227

Wheelwright, N.T. and Orians, G.H. (1982) Seed dispersal by animals: contrasts with pollen dispersal, problems of terminology, and constraints on coevolution. *American Naturalist* 119: 402–13. https://doi.org/10.1086/283918

Whitman, E., Parisien, M.-A., Thompson, D.K. and Flannigan, M.D. (2019) Short-interval wildfire and drought overwhelm boreal forest resilience. *Science Reports* 9: 18796. https://doi.org/10.1038/s41598-019-55036-7

Whittaker, R.H. (1977) Evolution of species diversity in land communities. In M.K. Hecht, W.C. Steere and B. Wallace, eds, *Evolutionary Biology*, pp. 1–67. New York: Plenum Press. https://doi.org/10.1007/978-1-4615-6953-4_1

Wiens, J.A. and Innis, G.S. (1974) Estimation of energy flow in bird communities: a population bioenergetics model. *Ecology* 55: 730–46. https://doi.org/10.2307/1934410

Willet, G. (1921) Bird notes from southeastern Alaska. *Condor* 23: 156–9. https://doi.org/10.2307/1362724

Williams-Guillén, K., Perfecto, I. and Vandermeer, J. (2008) Bats limit insects in a Neotropical agroforestry system. *Science* 320: 70. https://doi.org/10.1126/science.1152944

Willyard, A., Syring, J., Gernandt, D., Liston, A. and Cronn, R. (2007) Fossil calibration of molecular divergence infers a moderate mutation rate and recent radiations for *Pinus*. *Molecular Biology and Evolution* 24: 90–101. https://doi.org/10.1093/molbev/msl131

Wion, A.P., Weisberg, P.J., Pearse, I.S. and Redmond, M.D. (2020) Aridity drives spatiotemporal patterns of masting across the latitudinal range of a dryland conifer. *Ecography* 43: 569–80. https://doi.org/10.1111/ecog.04856

Wiseman, K.D., Greene, H.W., Koo, M.S. and Long, D.J. (2019) Feeding ecology of a generalist predator, the California Kingsnake (*Lampropeltis californiae*): why rare prey matter. *Herpetological Conservation and Biology* 14: 1–30.

Wyse, S.V., Brown, J.E. and Hulme, P.E. (2019) Seed release by a serotinous pine in the absence of fire: implications for invasion into temperate regions. *AoB PLANTS* 11: plz077. https://doi.org/10.1093/aobpla/plz077

Young, J.R., Braun, C.E., Oyler-McCance, S.J., Hupp, J.W. and Quinn, T.W. (2000) A new species of sage-grouse (Phasianidae: *Centrocercus*) from southwestern Colorado. *Wilson Bulletin* 112: 445–53. https://doi.org/10.1676/0043-5643(2000)112[0445:ANSOSG]2.0.CO;2

Young, J.R., Braun, C.E., Oyler-McCance, S.J., Aldridge, C.L., Magee, P.A. and Schroeder, M.A. (2020) Gunnison Sage-Grouse (*Centrocercus minimus*), version 1.0. In P.G. Rodewald, ed., *Birds of the World*. Cornell Lab of Ornithology, Ithaca, NY. https://doi.org/10.2173/bow.gusgro.01

Zachos, J., Pagani, M., Sloan, L., Thomas, E. and Billups, K. (2001) Trends, rhythms, and aberrations in global climate 65 Ma to present. *Science* 292: 686–93. https://doi.org/10.1126/science.1059412

Zangerl, A.R. and Berenbaum, M.R. (2003) Phenotype matching in wild parsnip and parsnip webworms: causes and consequences. *Evolution* 57: 806–15. https://doi.org/10.1111/j.0014-3820.2003.tb00292.x

Zhao, W., Gao, J., Hall, D., Andersson, B.A., Bruxaux, J., Tomlinson, K.W., Drouzas, A.D., Suyama, Y. and Wang, X. (2024) Evolutionary radiation of the Eurasian *Pinus* species under pervasive gene flow. *New Phytologist* 242: 2353–68. https://doi.org/10.1111/nph.19694

Ziswiler, V. (1965) Zur kenntnis des samenöffnens und der struktur des hörnernen gaumens bei körnerfressenden oscines. *Journal für Ornithologie* 106: 1–48. https://doi.org/10.1007/BF01670789

Zuccon, D., Prŷs-Jones, R., Rasmussen, P.C. and Ericson, P.G.P. (2012) The phylogenetic relationships and generic limits of finches (Fringillidae). *Molecular Phylogenetics and Evolution* 62: 581–96. https://doi.org/10.1016/j.ympev.2011.10.002

Zwolak, R., Celebias, P. and Bogdziewicz, M. (2022) Global patterns in the predator satiation effect of masting: a meta-analysis. *Proceedings of the National Academy of Sciences USA* 119: e2105655119. https://doi.org/10.1073/pnas.2105655119

Zylberberg, M., Van Hemert, C., Handel, C.M. and DeRisi, J.L. (2018) Avian keratin disorder of Alaska Black-capped Chickadees is associated with poecivirus infection. *Virology Journal* 15: 100. https://doi.org/10.1186/s12985-018-1008-5

Index